621.381528 L33T
LASTER, CLAY.
THYRISTOR THEORY AND
APPLICATION

2306703

**DO NOT REMOVE
CARDS FROM POCKET**

**ALLEN COUNTY PUBLIC LIBRARY
FORT WAYNE, INDIANA 46802**

You may return this book to any agency, branch,
or bookmobile of the Allen County Public Library.

THYRISTOR THEORY AND APPLICATION

CLAY LASTER

TAB BOOKS Inc.
Blue Ridge Summit, PA 17214

Allen County Public Library
Ft. Wayne, Indiana

**FIRST EDITION
FIRST PRINTING**

Copyright © 1986 by TAB BOOKS Inc.
Printed in the United States of America

Reproduction or publication of the content in any manner, without express permission of the publisher, is prohibited. No liability is assumed with respect to the use of the information herein.

Library of Congress Cataloging in Publication Data

Laster, Clay.
 Thyristor theory and application.

Includes index.
 1. Thyristors. 2. Thyristor control. I. Title.
TK7871.99.T5L38   1986   621.3815′287   85-27699
     ISBN 0-8306-0365-4
     ISBN 0-8306-0465-0 (pbk.)

Front cover photograph courtesy of Motorola Semiconductor Products Inc.

Contents

Preface — vii

Introduction — viii

1 Semiconductor Theory — 1
Basic Theory—PN Junctions—Transistors—Thyristors—How Thyristors Are Manufactured

2 The Silicon-Controlled Rectifier — 21
Basic SCR Operation—SCR Operating Characteristics—Gate Turn-On Circuits—Light Activated SCRs—Other Specialized SCR Devices—Introduction to Experiments—Experiment No. 2-1, SCR Characteristics—Experiment No. 2-2, SCR Ac Operation

3 The Triac — 71
Theory of Operation—Triac Gate Characteristics—Triac Turn-On Methods—Experiment 3-1, Triac Characteristics—Experiment 3-2, Triac Ac Operation

4 Other Thyristor Devices — 95
Shockley Diode—Diac—Silicon Controlled Switch—Gate Turn-Off Thyristor—Unijunction Transistor—Programmable Unijunction Transistor—Silicon Unilateral and Bilateral Switches

5 Planning the Thyristor Control System — 137
Initial Design—Environmental Considerations—Selecting the Proper Thyristor—Thermal Considerations—Radio Frequency Interference—Testing Thyristor Devices

6 Automation, Robots, and Thyristors   **159**
Automated Production and Processing—Microprocessors and Microcomputers—The Thyristor Connection

References   **177**

Index   **179**

Preface

THE PURPOSE OF THIS BOOK IS TO PROVIDE electronics engineers, technicians, and experimenters with a practical understanding of thyristor devices. The material—fundamentals, theory of operation, specifications, applications, and simple "laboratory-type" experiments—is organized primarily as a self-study text on thyristors. The final chapter, a *minicourse* in microprocessors and automation, is included to illustrate the interface between thyristors and the rapidly expanding field of industrial control systems and "robots."

The information for this book was obtained after a survey of all available thyristor devices. I wish to express my appreciation and thanks to the many commercial organizations who contributed technical data, specifications, and photographs. Without their splendid cooperation, this book on thyristors would not have been possible. Finally, I wish to thank the electronics faculty at San Antonio College for their constructive comments and assistance in developing technical information for this book.

Introduction

THYRISTORS, AN IMPORTANT PART OF THE semiconductor family, are generally defined as a group of solid-state devices used to control the switching of dc and ac power. Wherever ac or dc power is controlled or switched, thyristors are likely to be found. Anyone involved with the design, installation, or maintenance of electronic power control equipment needs to know what thyristors are and how they work.

This book explains how the various thyristor devices—such as the silicon controlled rectifier (SCR), triode ac semiconductor (triac), and the unijunction transistor (UJT)—are made and how they operate. Enough semiconductor theory is included to make the explanations understandable.

Chapter 1, "Semiconductor Theory," provides a brief review of fundamental theory of semiconductor devices and operation. This includes the characteristics of semiconductor materials, pn junctions, diodes, transistors, and thyristors. Chapters 2 through 4 cover the characteristics, theory of operation, and applications of specific thyristors, and simple "laboratory" experiments in Chapters 2 and 3 illustrate the principles of thyristor operation. This material is intended to provide practical, up-to-date information on the many different types of thyristors available on the market. Chapter 5 covers factors concerning the design, planning, installation, and maintenance of thyristor circuits for reliable operation. Such factors as thermal considerations, heat sink installation, and prevention of radio frequency interference are discussed in practical terms. Finally, Chapter 6 describes how thyristors are interfaced with digital control systems. This includes a general discussion of microprocessors and digital circuits used to control the switching of thyristors.

Chapter 1

Semiconductor Theory

THE FIELD OF ELECTRONICS TECHNOLOGY, which originated at the beginning of the twentieth century, has brought about fantastic changes in almost every phase of our lives. The invention of the Fleming valve (the diode vacuum tube) in 1902 and the DeForest triode tube in 1906 heralded the beginning of a revolution in technology unparalleled in human history. From these crude but important inventions sprang the fields of electronic communications, industrial and medical electronics, digital computers, the microprocessor (or "computer-on-a-chip"), and a host of other developments.

At the time when the diode tube was invented, crystal detectors were used to detect or rectify radio waves. The crystal detectors represented the first practical group of semiconductor devices. Unfortunately, the theory behind these devices was not fully developed until some 40 years later. Despite this lack of theoretical knowledge, selenium and other diode rectifiers were developed in the 1930s. These devices were fabricated primarily on a "cut-and-try" basis.

The big breakthrough in semiconductor physics came about in the late 1940s. Scientists at the Bell Telephone Laboratories developed the theory of semiconductor action and built the first transistor. In 1956, Bell Telephone Laboratories scored another significant first when they developed the silicon-controlled rectifier (SCR). By 1957, the General Electric Company had produced the first commercial silicon-controlled rectifiers. Since that time, a wide range of SCRs and related thyristor devices have been developed to meet virtually any power switching or control application.

The development of integrated circuit (IC) technology in 1958 by the Texas Instruments company permitted the fabrication of complete electronic circuits on tiny chips of semiconductor material. This resulted in significantly lower production costs and greater reliability. Today, ICs are available for almost any electronic requirement.

BASIC THEORY

In order to understand electrical current, it is first necessary to understand the nature of matter.

All matter is made up of submicroscopic molecules. Molecules are made up of atoms, and atoms in turn are made up of particles called *electrons, protons* and *neutrons*. Scientists tell us today that there are many more types of particles within the atom. The theory of electrical current however, can be explained using only these three basic particles. Each of the chemical elements found in nature is made up of atoms that contain a unique combination of the three basic particles. To date, some 103 kinds of atoms (103 different chemical elements) have been discovered.

Atomic Structure

The structure of an atom is illustrated in Fig. 1-1A. Protons and neutrons, the heavier particles, make up the nucleus. Electrons, which are lighter particles, rotate around the nucleus in specified orbits. In a general sense, the sun and its orbiting planets are similar to a giant atom.

Figure 1-1 also shows the configuration of the hydrogen, copper, and silicon atoms. The simplest atom, hydrogen, consists of one electron rotating around a nucleus containing one proton (Fig. 1-1B). The copper atom (Fig. 1-1C), one of the most efficient conductors of electricity, has a nucleus of 29 protons and 34 neutrons, and 29 orbiting electrons. Silicon atoms (Fig. 1-1D), the major semiconductor material, possesses 14 protons and 14 neutrons in the nucleus, and 14 orbiting electrons.

Positive and Negative Charges

An *electron* is a tiny particle that possesses a unit negative electrical charge. When electrons are

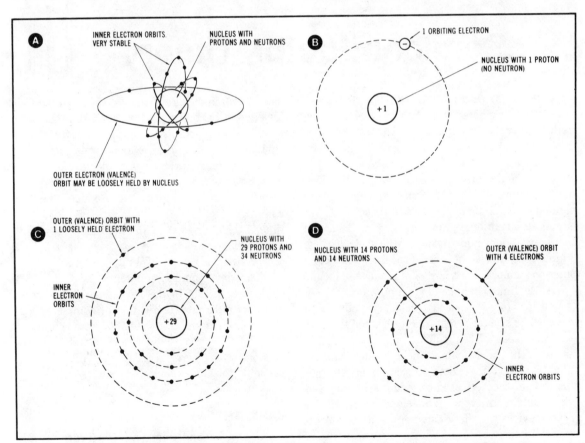

Fig. 1-1. Atomic configurations. (A) Generalized model. (B) Hydrogen atom. (C) Copper atom. (D) Silicon atom.

dislodged from the outer orbits of their parent atoms, they are free to move. Free electrons are sometimes called negative ions or mobile electrons. Since moving electrons are moving charges, and since charge in motion is current, the motion of free electrons constitutes an *electrical current*. This leads to the basic definition of one *ampere* of current: when 6.28×10^{18} electrons pass a plane in a conductor in a period of one second, the current is defined as one ampere.

The *proton* possesses a unit positive electrical charge. You will note in Fig. 1-1, each atom has an equal number of electrons and protons. Such atoms are referred to as neutral atoms since they possess a zero net electrical charge. The negative charge of one electron cancels the positive charge of one proton.

Neutrons have no electrical charge; they contribute only to the mass of the nucleus. In general, the smaller atoms contain approximately equal numbers of neutrons and protons. Neutrons outnumber the protons in the larger, more complex atoms.

Conductors, Semiconductors, and Insulators

A material may be classified as a conductor, semiconductor, or insulator, depending on the ability of the material to produce free electrons. A large number of free electrons allows a greater electrical current to be conducted within the material.

Conductors such as silver and copper possess many free electrons. The atoms in these materials possess a few loosely bound electrons in their outer orbits. Energy, in the form of heat, can dislodge these outer orbit electrons, causing them to drift within the material. Copper and silver atoms, for example, possess only one electron in the outer orbit. At normal room temperature, a tiny piece of silver or copper wire produces billions of free electrons. An atom which loses an outer-orbit electron will have a unit positive charge; such an atom is called a positive ion.

Insulators are materials which do not conduct electrical current. Glass, ceramics, and plastics are examples of insulating materials. The atoms that make up these materials do not produce free electrons under normal conditions. All electrons within each atom are tightly bound to the nucleus. The absence of free electrons means that no electrical charge can be conducted through the material. Outer-orbit electrons associated with these atoms can only be dislodged by the presence of an extremely strong electric field. This process results in an avalanche of high-density current which usually causes physical damage to the insulating material. This action is referred to as *breakdown*. We will also use this characteristic in describing the action of semiconductor devices.

Semiconductor materials fall into an in-between category; they act as neither good conductors nor good insulators. Although many materials exhibit this characteristic, silicon and germanium are the most widely used semiconductors.

Each semiconductor atom contains four electrons in its outer orbit. These electrons are more tightly bound to the nucleus than the outer-orbit electrons of conductor atoms. Hence, more energy is required to dislodge electrons from the semiconductor atoms. At normal room temperature, only a relative few free electrons are available for conducting current. For this reason, semiconductor materials *in pure form* are seldom used in the fabrication of solid-state devices.

Electron and Hole Current

The flow of electrical current in a conductor is made possible due to the abundance of free electrons within the conductor. When an electrical potential, or *voltage*, is impressed across the conductor, the free electrons drift toward the positive pole of the voltage source. The simple battery-conductor circuit in Fig. 1-2A illustrates this action.

The *conductivity* of a given material, such as copper, is directly related to the number of free electrons indicates a higher conductivity. *Resistance,* given size of conductor, a greater number of free electrons indicate a higher conductivity. *Resistance,* the inverse of conductivity, is defined as the opposition to current flow in a circuit. The unit of resistance is the *ohm*.

The relationship between resistance, current, and voltage can be expressed as follows: when 1

3

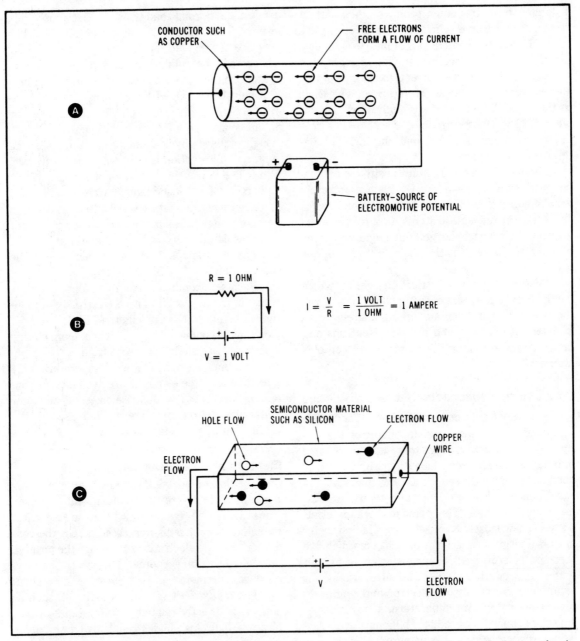

Fig. 1-2. Electrical current in conductors and semiconductors. (A) Current in a conductor. (B) Schematic diagram of a circuit. (C) Flow of electrons and holes in a semiconductor.

volt (V) of electrical potential is impressed across a conductor with a resistance of 1 ohm, 1 ampere (A) of current will be produced in the conductor (Fig. 1-2B). You may recognize this as a special case of Ohm's Law:

$$I = \frac{V}{R} \qquad \textbf{(Eq. 1-1)}$$

where,
I is the current in amperes,
V is the voltage in volts,
R is the resistance in ohms.

In order to explain current in a semiconductor material, we will introduce the concept of *hole current*. When a semiconductor atom loses an outer-orbit electron, a hole is created in this orbit. The hole represents a positive charge carrier that can drift from atom to atom. If a battery is connected across a block of semiconductor material, the *holes* will drift toward the *negative* pole of the battery. At the same time, free *electrons* within the material will drift toward the *positive* pole. The flow of holes and free electrons in a semiconductor is shown in Fig. 1-2C.

The ability of holes to drift within a semiconductor is due to the crystalline structure of the material. The atoms of most solid matter are locked together in a crystalline form, a sort of atomic lattice, as shown in Fig. 1-3. Adjacent atoms share their outer-orbit electrons with each other. Thus an atom that has lost an electron possesses a hole that can be passed to adjacent atoms.

Figure 1-3 illustrates the crystalline structure of pure silicon or germanium. Note that in order to simplify this model, the inner electron orbits are not shown. The few free electrons created in *pure* semiconductor material are not adequate for conducting a useful electrical current in electronic circuits.

Although pure semiconductor materials are not practical for use in electrical circuits, we can increase the number of free electrons and holes by adding, or *doping*, impurities into the semiconductor. The modified semiconductor material then exhibits properties that are useful in solid-state devices such as thyristors, transistors, and diodes.

N-Type Semiconductors

The addition, or doping, of small amounts of pentavalent atoms into a semiconductor material will increase the number of free electrons by a substantial amount. The term *pentavalent* defines a class of atoms that have five electrons in their outer orbits. Chemists and semiconductor physicists refer to the outer-orbit electrons as valence electrons. Antimony, arsenic, and phosphorus are typical of the pentavalent materials used as doping agents in semiconductors.

The doped, or modified, material is called negative, or *N-type*, semiconductor due to the large numbers of free electrons available. Each pen-

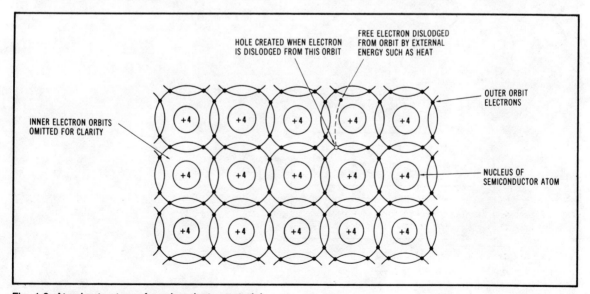

Fig. 1-3. Atomic structure of semiconductor material.

tavalent atom is locked into the crystalline structure, its outer-orbit electrons merging with the outer orbits of neighboring semiconductor atoms. Figure 1-4A illustrates silicon material doped with arsenic atoms. This concept is applicable to any semiconductor doped with a pentavalent material.

The arsenic atom can contribute only four of its five valence atoms to the interlocking outer orbits of adjacent atoms. Thus the fifth electron of the arsenic atom is free to drift within the semiconductor material. The arsenic atom is called a *donor atom*, since it donates an additional free electron to the material. A small amount of donor impurity, approximately one donor atom for each one million semiconductor atoms, is sufficient to increase the current carrying capacity to usable levels. Figure 1-4B shows a block of N-type semiconductor connected to a voltage source. The free electrons liberated by the doping process are called *majority carriers* because their motion accounts for most of the current flowing through the material.

A very small percentage of the total current being conducted in N-type material is due to small number of holes created by thermal energy. These holes are referred to as *minority carriers*. The holes can be visualized as carrying a positive charge.

P-Type Semiconductors

The number of holes within a semiconductor can be substantially increased by the addition of a *trivalent* impurity such as gallium or boron. These are atoms which possess three electrons in their outer orbits. When doped into the crystalline structure of a semiconductor, the trivalent atoms have

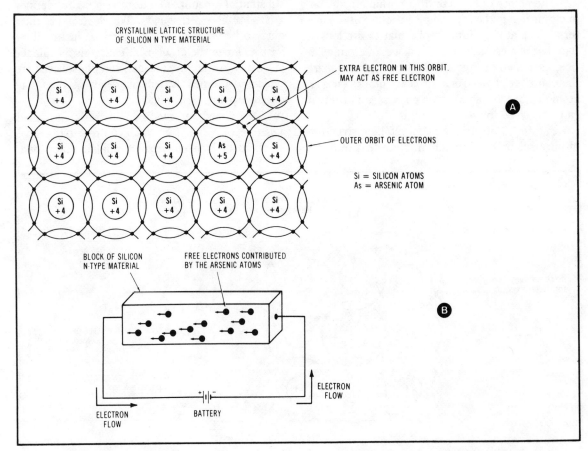

Fig. 1-4. Typical N-type semiconductor material. (A) Model of silicon doped with arsenic. (B) Electron flow.

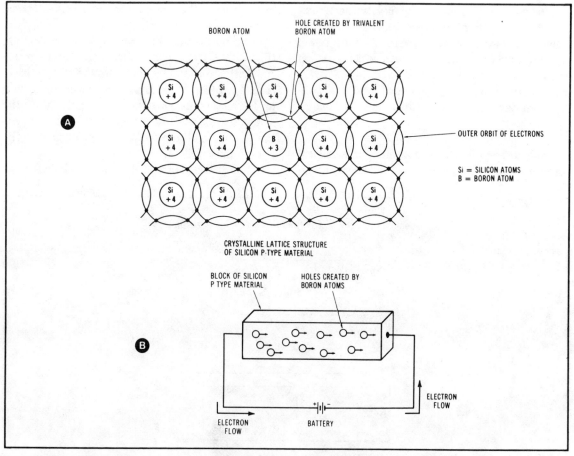

Fig. 1-5. Typical P-type semiconductor material. (A) Model of silicon doped with boron. (B) Hole flow.

only three electrons to contribute to the interlocking orbits of adjacent atoms. The absence of an electron creates a hole which is free to drift into other interlocking orbits. Figure 1-5A represents silicon semiconductor material doped with boron atoms. The trivalent atoms are called *acceptor* atoms since they can accept an additional electron in the orbit that interlocks with semiconductor atoms.

This modified material is known as positive, or *P-type*, semiconductor since a large number of positive charge carriers are created. As in the N-type semiconductor, only a small amount of trivalent impurity is required to produce P-type semiconductor. The addition of approximately one acceptor atom per million semiconductor atoms is adequate to produce usable currents in P-type material. Figure 1-5B illustrates hole current in P-type semiconductor. *Holes* created by the doping process are called *majority carriers*. The *minority carriers* in P-type material are small numbers of free *electrons* produced by thermal energy.

PN JUNCTIONS

The increased conductivity characteristics of P-type and N-type semiconductor materials are roughly equivalent to those of carbon resistors. Materials such as copper or silver make much more efficient conductors of electrical current. Why then are P-type and N-type semiconductor materials so important to the operation of solid state devices?

The answer to this all important question lies in the *pn junction*. When tiny blocks of P-type and

N-type materials are joined together, we have a device that exhibits unique electrical characteristics. The pn junction allows current to be conducted in only one direction; current in the opposite direction is virtually cut off. One practical application of this characteristic is the pn-junction diode, a two-terminal device.

Unbiased PN Junctions

When the pn junction is formed, an electrical force or potential is created across the junction. Figure 1-6 shows the resulting action of this electrical force.

During the initial formation of the pn junction (Fig. 1-6A), free electrons from the N-type material drift across the junction into the P-type material (Fig. 1-6B). These free electrons combine with (or fall into) the holes in the P-type material. At the same time, holes in the P-type material drift across the junction and combine with free electrons in the N-type material.

The diffusion of free electrons and holes across

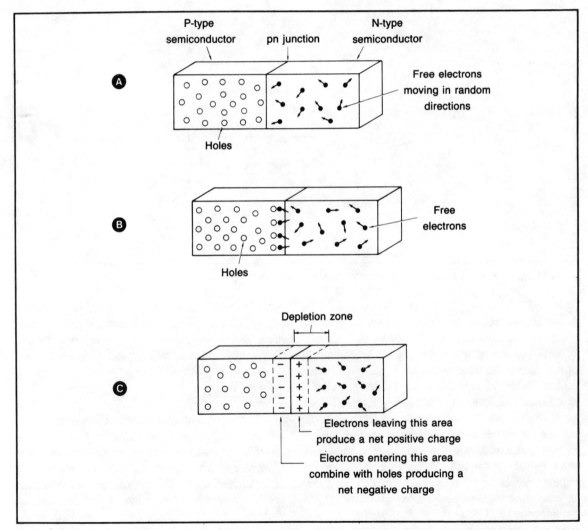

Fig. 1-6. Electrical characteristics of the pn junction. (A) Initial condition. (B) Free electrons crossing junction. (C) Final condition.

the pn junction creates a net negative charge zone on the P-side and a positive charge zone on the N-side (Fig. 1-6C). This diffusion and subsequent buildup of electrical charge continues until the resulting electrical force is sufficient to prevent further diffusion of electrons and holes. The negative charge zone created in the P-type material repels any additional free electrons trying to cross the junction. This positive-negative charge region is known as the *junction barrier* or *depletion zone*.

The junction barrier exists for only a very small distance on each side of the pn junction. The positive and negative space-charge regions form a potential difference which acts as an internal battery. Figure 1-6C illustrates this concept. The barrier potential cannot be measured directly with a voltmeter since the net electrical charge across the pn block of material is zero. Indirect voltage measurements, however, will indicate a barrier potential of approximately 0.3 volt for germanium and approximately 0.7 volt for silicon pn junctions.

Biased PN Junctions

Diodes, transistors, and most thyristors employ pn junctions to accomplish their required operations. When connected into an electrical circuit with an external power source, these devices perform such functions as rectifying as ac current, amplifying an ac signal, or shaping signal waveforms.

The pn-junction diode, a two-layer pn-junction device, is illustrated in Fig. 1-7. The diode allows current to be conducted in one direction, while the electrical properties of the pn junction prevent current flow in the opposite direction. Connecting the diode in a dc circuit will help to explain this unique phenomenon.

Forward Biasing. If a pn-junction diode is connected in the dc circuit shown in Fig. 1-8A, the diode permits current to pass. The diode is said to be forward biased, since the anode is more positive than the cathode.

Note that the positive pole of the battery is connected to the P-type portion of the diode and the negative pole is connected to the N-type material through series resistor R. Resistor R is placed in the circuit merely to limit the current through the

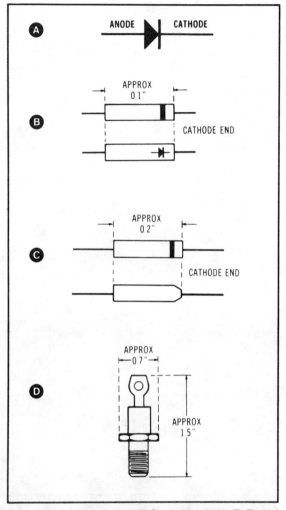

Fig. 1-7. Pn junction diodes. (A) Symbol for diode. (B) Typical general-purpose signal diode. (C) Typical low-current rectifier. (D) A 35-ampere silicon rectifier.

diode. Without this resistor, excessive current would quickly destroy the diode.

Current in this circuit is due to the following factors. The larger numbers of free electrons in the N-type material are attracted toward the positive potential of the battery. If this force of attraction is greater than the junction barrier potential, the free electrons will drift across the junction. You can equate this force of attraction to the potential difference (voltage) of the external power source. For example, if a silicon diode is used in this circuit,

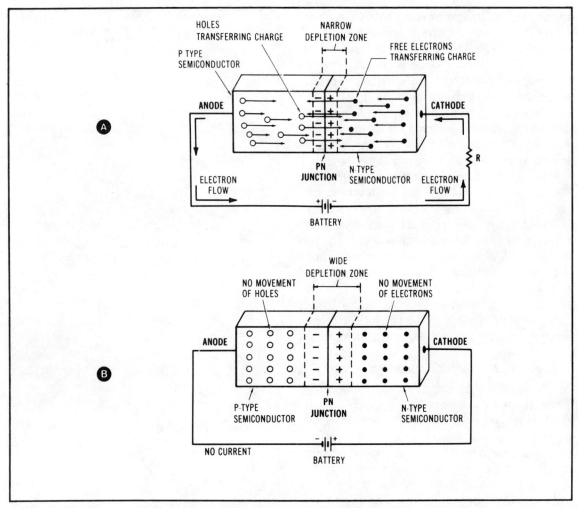

Fig. 1-8. Forward and reverse biasing for pn junction diodes. (A) Forward bias. (B) Reverse bias.

the battery potential must be slightly more than 0.7 V in order to overcome the 0.7-V barrier potential of the silicon diode. Similarly, the forward bias voltage for germanium pn-junction diodes must be slightly greater than 0.3 V in order to permit a flow of free electrons across the junction.

If you increase the forward bias voltage, more electrons flow across the pn junction. The large number of electrons in the vicinity of the junction reduces the size of the depletion zone. Thus, any increase in forward bias voltage results in a corresponding increase in current through the diode.

Reverse Biasing. If the battery connections to the diode are reversed, the current will be virtually zero. This is illustrated in Fig. 1-8B. The new circuit results in the cathode being at a positive potential with respect to the anode. This is referred to as reverse biasing.

The free electrons in the N-type material are attracted toward the positive potential of the cathode terminal. At the same time, holes in the P-type material drift toward the negative potential on the anode terminal. These actions expand the depletion zone, and the internal barrier potential assumes a value equal to that of the external applied voltage.

Leakage Current. A small leakage current

is present in any reverse-biased diode. The value of the leakage current depends primarily upon the type of material used in fabricating the diode and upon the operating temperature. An increase in operating temperature results in a corresponding increase of leakage current.

The leakage current of silicon diodes at normal operating temperatures is so small that it can be neglected for most applications. For example, the typical leakage current for a 1N4004 silicon rectifier diode is on the order of 50 nanoamperes (nA) when a reverse bias of 400 V is applied across the diode.

Germanium diodes exhibit much higher leakage currents. The 1N270 germanium diode, for example, shows a typical leakage current of approximately 100 microamperes (μA) for an applied reverse bias of 80 volts.

PN-Junction Diode Operating Characteristics

The voltage-current characteristic curve of a diode indicates how the diode will operate in a circuit. For ease of comparison, two voltage-current (VI) curves are shown in Fig. 1-9. The solid-line curve represents the operating characteristics of the 1N3063 silicon high-speed switching diode, and the dash-line curve shows the operating characteristics of the 1N276 germanium high-conductance diode. Note that the scales have been expanded where necessary to show the characteristics of the diodes. Also, *some* silicon diodes *may* have a lower breakdown voltage when compared to a particular germanium diode.

Forward current in a diode increases very slowly at low values of forward bias voltage. This is illustrated in the upper right quadrant of the graph. However, when the applied forward voltage across the diode is increased beyond the value of the junction barrier potential, the current through the diode increases rapidly. If the forward bias voltage is increased beyond the manufacturer's spec-

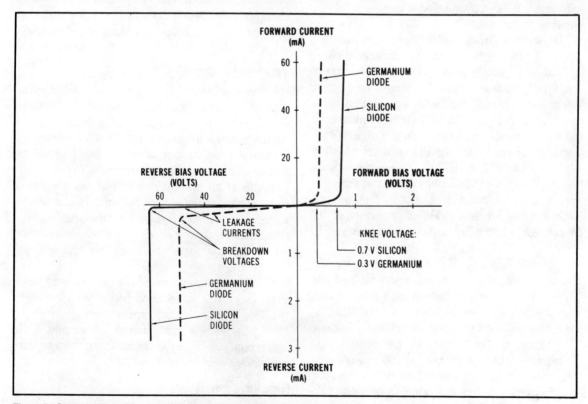

Fig. 1-9. Characteristic VI curves for typical silicon and germanium diodes.

ifications, the diode could be destroyed by the resulting excessive forward current.

When the diode is reversed biased, as in the lower left quadrant of Fig. 1-9, the reverse current increases very slowly until the breakdown voltage is reached. At this point, the reverse current increases so rapidly that the diode may be quickly destroyed. Manufacturers usually specify the maximum safe reverse bias voltage as the *peak inverse voltage* (PIV).

Sometimes the breakdown voltage is referred to as the avalanche region or *zener* voltage. A special class of diodes, such as the zener (voltage-reference) diodes, are operated in the breakdown-voltage region. These devices are heavily doped to permit large values of reverse current without overheating.

Types of Semiconductor Diodes

Diodes are manufactured for a variety of applications. We can divide diodes into the following general categories.

Rectifier Diodes. Rectifier diodes are used primarily for high-current applications such as changing ac to dc in power supplies. The current-carrying capacities of rectifier diodes range from about one ampere to hundreds of amperes. For example, the 1N4001 to 1N4007 series of silicon rectifiers are rated for one ampere of average forward rectifier current. Having PIV ratings of 50 V (1N4001) to 1000 V (1N4007), these devices are used in low-power applications—such as am or fm radios. Large silicon diodes, such as the Motorola 1239 and the Sylvania ECG6353, are rated at 300 A average rectified forward current and have PIV ratings of 600 V.

Signal Diodes. Signal diodes are generally characterized as low-current, fast switching devices. They are used in high-speed digital circuits, high frequency mixers and detectors, and other circuits that require high-speed switching operations. Fast switching diodes, such as the 1N3485, have a switching time of approximately 50 nanoseconds. Manufacturers rate the switching time in terms of reverse recovery time (t_{rr}), which is defined as the time required for the diode to recover from forward current flow and begin to block reverse current.

Zener Diodes. Zener diodes are heavily doped diodes designed to operate in the reverse-biased breakdown-voltage region. Zener diodes are manufactured for operation at various specified breakdown voltages. This characteristic makes them useful as voltage regulators, waveform clippers, and for other related functions. Zener diodes are available with breakdown voltages from about 1.8 V to 200 V. Power ratings of these devices range from about 250 milliwatts (mW) to 50 watts (W).

Special-Purpose Diodes. The unique characteristics of semiconductor materials have allowed the development of many specialized types of diodes. These include the *varactor* (variable-capacitance diode); *hot carrier* (*Schotty*) diodes for vhf, uhf, and microwave mixer and detector circuits; *pin-switching* diodes for vhf and uhf rf switching circuits; and *tunnel* diodes, used in oscillator and other circuits that require a negative resistance characteristic. Space does not permit a detailed description of these devices. Many excellent texts are available for further research in this area.

TRANSISTORS

Semiconductor pn junctions can be arranged in a three-layer configuration, either npn or pnp, to produce the *bipolar transistor*. This three-terminal device acts as a current amplifier or electronic switch. A small change in input current produces a large change in output current.

Almost all present day transistors are made from silicon semiconductor material. In addition to excellent reliability and high-temperature operation, silicon possesses electrical characteristics which are superior to those of other semiconductor materials.

Other configurations of the pn junctions have been employed to produce *field-effect transistors* (FETs). These devices are essentially voltage amplifiers since a small change in the input voltage produces a large change in output voltage, current, or power.

Bipolar Transistors

Bipolar, or junction, transistors are produced

in two basic types: npn or pnp structures. As shown in Fig. 1-10A, the npn transistor consists of outer blocks of heavily doped N-type material separated by a thin section of lightly doped P-type material. The pnp transistor (Fig. 1-10B) employs the reverse of this construction, consisting of two blocks of heavily doped P-type material separated by a thin section of lightly doped N-type material. Note that each type of transistor contains two separate pn junctions. The thin inner sections, together with the proper biasing of each pn junction, provide the key to transistor operation.

Biasing the NPN Transistor. Figure 1-11 shows the three proper biasing circuits for the npn transistor. Note that in each case, the base-emitter pn junction is forward biased while the base-collector pn junction is reverse biased. This combination of forward and reverse biasing results in a large current from emitter to collector. However, the base current (I_B) is found to be very small, about 5 percent of less of the emitter current (I_E).

The basis for this unusual current relationship may be visualized using the *common base* circuit in Fig. 1-11A to analyze the transistor action. Free electrons from the emitter are swept into the thin base region by the forward biased emitter-base pn junction. A few of the electrons will combine with the limited number of holes in the P-type base material. This allows a small current to flow out of the base terminal toward the positive pole of the V_{EE} battery. However, most of the free electrons entering the narrow base region will drift into the N-type collector region. The positive potential on the collector attracts the electrons and causes current flow toward the positive pole of the V_{cc} battery. The resulting current flow is conducted back toward the emitter terminal. You will find a simple but important relationship here—I_E is equal to the sum of I_B and I_C.

An important aspect of this circuit action is that a small change in the base current (I_B) produces a large change in collector current (I_C). For example,

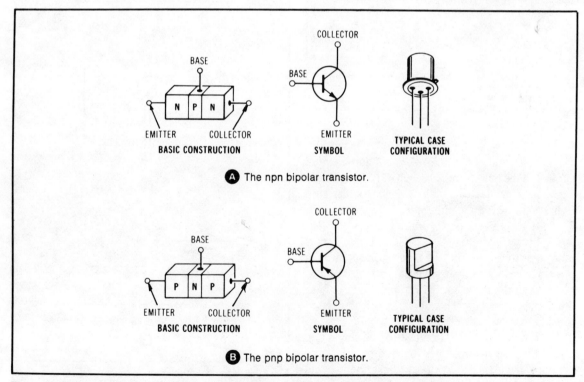

Fig. 1-10. Bipolar transistors. (A) The npn bipolar transistor. (B) The pnp bipolar transistor.

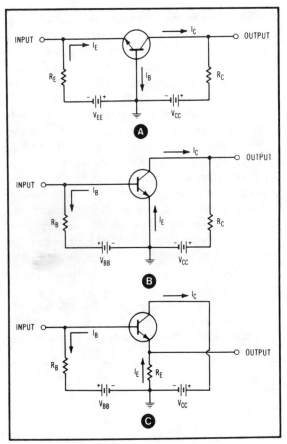

Fig. 1-11. Biasing circuits for npn transistors. (A) Common base. (B) Common emitter. (C) Common collector.

junction and reverse biasing of the base-collector pn junction (Fig. 1-12). The pnp transistor works in the following manner. The excess holes in the emitter flow into the thin, lightly doped base region. Only a few of these holes (about five percent or less) combine with free electrons in the base region. This results in a small base current. Most of the holes drift into the collector region and are attracted to the negative potential on the collector terminal. Free electrons supplied by the conductor connected to the collector terminal combine with the holes leaving the collector region.

The pnp transistor can also be operated in the three basic circuit configurations: common emitter, common base, and common collector. This is il-

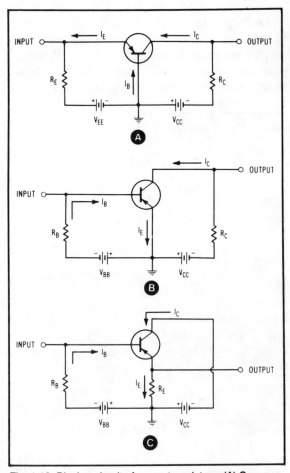

we can use a transistor in the arrangement of Fig. 1-11B as an audio amplifier. A small audio signal applied to the base terminal results in a large audio signal in the collector circuit. This particular configuration is known as the *common emitter* circuit. The other transistor configuration, *common collector*, is illustrated in Fig. 1-11C. There is one important point to remember: the transistor operates in the same manner for any configuration. The differences in the three configurations relate only to input and output signal connections and associated impedance levels.

Biasing the PNP Transistor. Using a pnp transistor, and reversing the battery polarity connections from those previously described, again results in forward biasing of the base-emitter pn

Fig. 1-12. Biasing circuits for pnp transistors. (A) Common base. (B) Common emitter. (C) Common collector.

lustrated in Fig. 1-12. Note again that the power supply voltage connections and resulting current flow are reversed in comparison with npn transistor circuits.

Pnp transistors are generally slower to react in operation than npn transistors because hole mobility is slightly slower than electron mobility. Also, since npn transistors usually cost less to produce, the npn transistor is the most popular type available today.

Field Effect Transistors

The *field-effect transistor* (FET) is a voltage-controlled device; that is, a small change in the input voltage produces a large change in output voltage, current, or power. Operation of the FET involves an electric field which controls the flow of charge through the device. In contrast to the low input impedance of bipolar, or junction, transistors, FETs exhibit an extremely high input impedance on the order of thousands of megohms.

There are two categories of FETs: *junction FETs* (JFETs), and *metal-oxide-semiconductor FETs* (MOSFETs). Symbols for these devices are given in Fig. 1-13. The *source, drain,* and *gate* are analogous to the *emitter, collector,* and *base* sections of bipolar transistors. The terms *n-channel* and *p-channel* refer to the type of material to which the source and drain leads are attached.

THYRISTORS

Further research involving semiconductor pn junctions paved the way for development of multilayer pnpn devices known as *thyristors*. The term thyristor applies to a broad family of solid-state devices designed for electronic switching and control of ac and dc power. The older vacuum tube device for switching power circuits was called thyratron—in a broad sense, you can equate thyristor to *transistorized thyratron*. The *silicon-controlled rectifier* (SCR), for example, is a thyristor for controlling current in one direction and blocking current in the reverse direction. Other thyristors, such as the *triac*, can be used to switch current in either direction. As you have probably

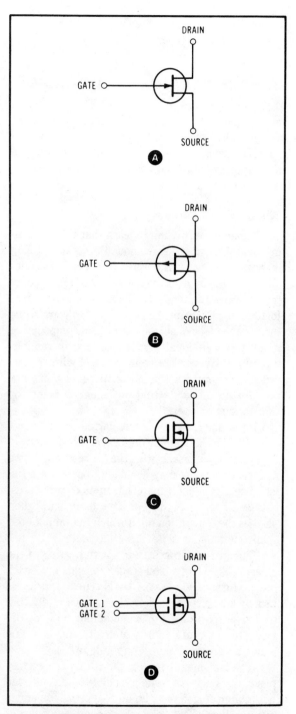

Fig. 1-13. Common types of field effect transistors. (A) N-channel JFET. (B) P-channel JFET. (C) N-channel MOSFET. (D) Dual gate N-channel MOSFET.

concluded, the semiconductor thyristor has made the older thyratron tube obsolete.

HOW THYRISTORS ARE MANUFACTURED

Thyristors are manufactured by a variety of highly specialized processes involving some six or seven individual steps. During each stage of manufacturing, strict quality control and automated testing are required to ensure a usable product.

Silicon From Ordinary Sand

It may be surprising to learn that the basic ingredient for most semiconductor devices being manufactured today is derived from ordinary sand. The first step in producing any silicon semiconductor device is refining pure silicon from sand. Unfortunately, the refined silicon is in a *polycrystalline* form; each piece of silicon material is composed of many tiny individual crystals aligned in random directions. Even when contaminated with N-type or P-type impurities, this silicon material would not be capable of conducting sufficient current for semiconductor operation.

Pure silicon can be converted to a *monocrystalline* form, or single uniform crystal structure, by a growth process. During the process, an N-type or P-type impurity may be added to produce a finished semiconductor material. In most processes, N-type silicon is produced. This material is then used as the foundation upon which the unmounted semiconductor device will be fabricated.

The growth process starts with heating of the refined silicon to a molten, or liquid, state in a crucible (a furnace-like vat). An N-type impurity is usually added to the molten silicon in carefully controlled proportions. For some semiconductor devices, the ratio of impurity atoms to silicon atoms is on the order of one to one million.

The growing process starts with lowering a tiny piece, or "seed," of monocrystalline silicon into the hot molten silicon. The liquid N-type silicon starts to crystallize on the solid silicon seed, and a larger mass of solid N-type silicon is formed. Then the silicon seed is gradually withdrawn from the crucible of molten silicon. At the same time, the silicon seed structure and the crucible are rotated in opposite directions to ensure a uniform structure of grown N-type silicon material. The end result is a cylindrical rod of N-type silicon material with a diameter of about two inches or more.

Each silicon rod is carefully sliced into many thin discs, or wafers. These wafers serve as the foundation for manufacturing semiconductor devices. For example, one wafer can be used for making one large high-current SCR, about 100 small SCRs, or many integrated circuits.

Forming PN Junctions

There are many manufacturing techniques by which pn junctions can be formed. The more common processes are called *grown junction, alloy, diffusion,* and *epitaxial.* The *grown junction* is produced by the alternate addition of P-type and N-type impurities to molten pure semiconductor material such as silicon. This creates a semiconductor *ingot* which possesses a number of alternate layers of P-type and N-type materials. The ingot can be sliced into thin wafers of N-type material joined with P-type material to form pn junctions.

The *alloy* process involves placing small balls of a P-type impurity, such as aluminum, on the N-type silicon wafer. Then the two materials are heated in a furnace until the aluminum melts and mixes with the silicon. The resulting alloy of aluminum and silicon forms a P-type silicon material and the desired pn junction. Although this process was used extensively in manufacturing germanium transistors, it is seldom used in fabricating silicon devices.

The *diffusion* process is similar to the alloy process in that a P-type impurity is diffused into one side of an N-type silicon wafer. Either solid or gaseous diffusion is used to convert a thin layer of N-type silicon into P-type silicon. The process is carried out in a controlled heated environment to speed up the diffusion activity. The first diffusion results in creating a single pn junction. If desired, the wafer can be sliced into many small pn sections and assembled as pn junction diodes.

Another common process, the *epitaxial,* or

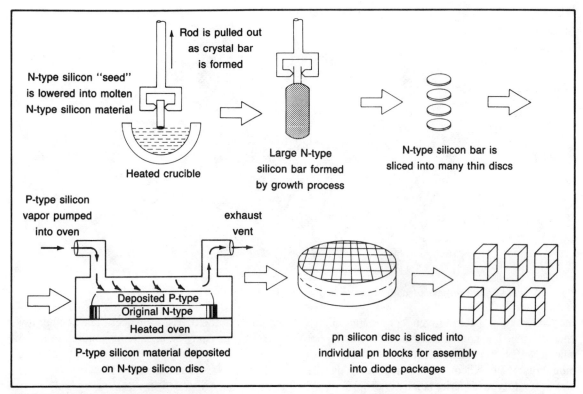

Fig. 1-14. Manufacture of pn silicon wafers.

vapor depositing, process, is used to manufacture pn junctions. Figure 1-14 illustrates the manufacture of pn-junction diodes by the epitaxial deposition of P-type silicon material over N-type silicon discs. In many instances, a combination of processes is used during the manufacturing process.

In order to create additional pn junctions on the same wafer, additional diffusions of N-type and P-type impurities are employed. In some cases, photoetching methods are used to control the specific location of pn junctions. Finally, the wafer is sliced and lapped into the required geometric shapes. Many types of thyristors are manufactured by the diffusion process.

The third basic process for forming pn junctions is that of *epitaxial* growth. As in the alloy and diffusion processes, an N-type wafer usually serves as the foundation. However, additional layers of N-type and P-type silicon material are "grown" on one side of the wafer. This is accomplished by placing

Fig. 1-15. Fabrication of thyristors on silicon wafers (courtesy of Teccor Electronics Inc.).

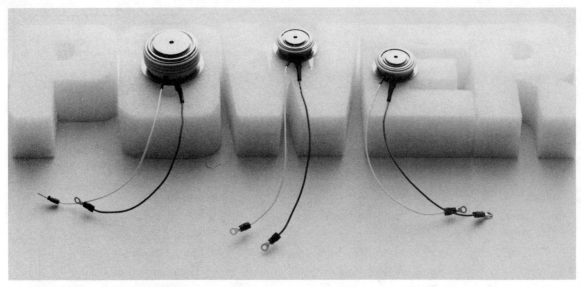
Fig. 1-16. Some typical thyristors (courtesy of Westinghouse Electric Corp.).

the wafer in a heated oven containing silicon chloride gas. The chemical action between the hot silicon wafer and the gas causes a thin layer of monocrystalline silicon to be deposited on the wafer. If a P-type impurity (such as boron) is mixed with the silicon chloride gas, the result is a P-type silicon layer epitaxially grown on the N-type wafer. Thus, a pn junction is formed on the silicon wafer. Figure 1-15 shows the wafers during one phase of manufacture.

Final Assembly and Packaging

The unmounted thyristor device is shaped to the required dimensions and cleaned to remove any impurities caused by the manufacturing processes. In many cases, each unmounted thyristor is fully tested before final assembly. An alternate test method is to select representative samples for test and evaluation.

Many thyristors, particularly the high-current types, are mounted on a metallic base such as molybdenum or tungsten. This base, or *substrate* as it is sometimes called, serves to protect the fragile silicon device from vibration or mechanical shock.

A wide variety of packages is used for the final assembly of the thyristor devices. The smaller low-current types may be installed in small metal cans such as the TO-3 and TO-5 cans, or in plastic cases such as the TO-98 and TO-220. The large thyristors are generally installed in metal cases, some with integral heat sinks. In many instances, the high-current thyristor is encapsulated in a disc-type package that has flat metallic sides for double-side heat-sinking. As will be seen later, adequate heat dispersal is a prime consideration in the installation and use of power thyristors.

Thyristors are available in many sizes and varying current and voltage ratings. For example, the current-carrying capacities of low power SCRs and triacs mounted in small plastic packages range up to about 25 A. Medium-power SCRs and triacs installed in metal cases control current up to about 50 A. High-power SCRs are available for handling currents as high as 2000 or more. The reverse (blocking) voltage ratings of SCRs generally range from about 15 to 800 V. Figure 1-16 shows some of the typical thyristor devices available from electronics parts suppliers.

A summary of the thyristor family, along with electrical symbols, is given in Fig. 1-17. We will cover these devices in greater detail beginning with Chapter 2.

SCR—Silicon Controlled Rectifier

The SCR, probably the most important member of the thyristor family, is basically a half-wave silicon rectifier with a third terminal, called the gate. An input gate signal causes the SCR to turn on, allowing forward current conduction. Like the diode, the SCR blocks conduction in the reverse direction.

A limitation of the SCR is the inability of the gate to turn off forward current. Once the SCR is turned on, the only way to turn it off is to reduce the forward current to a value less than the minimum holding current.

Turn-on time for SCRs is very fast, on the order of 0.1 to 1.0 microsecond. However, turn-off time is much slower, typically 5 to 30 microseconds.

Triac—Bidirectional Triode Thyristor

The triac, a three-terminal device, has a gate terminal for controlling the flow of current in either direction. A positive signal at the gate lets the triac "fire," or conduct, in one direction; a negative signal fires the triac for conduction in the opposite direction.

Triac turn-on and turn-off times are usually slower than those of SCRs. The turn-on time for triacs, for example, is about 1 to 5 microseconds.

The triac provides full-wave control of ac power for such applications as varying the speed of motors, light-dimming circuits, and temperature control. Note that two SCRs would be required to perform the function of one triac.

Diac—Bidirectional Trigger Diode

Diacs, sometimes called trigger diodes, are two-terminal devices normally used to control the turn-on of triacs. The diac conducts in either direction when the applied potential reaches about 27 to 37 volts.

Most diacs are low-power devices having a power dissipation of one watt or less with peak current limited to about 200 milliamperes.

SCS—Silicon Controlled Switch

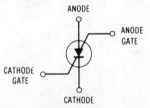

The SCS is a four-terminal device similar in construction and performance to the SCR. However, unlike the SCR, the SCS can be turned off by appropriate gate voltage. Either of the two gate terminals can be used to turn on or turn off the flow of current from anode to cathode. Also, the SCS exhibits a fast turn-off time, typically 1 to 10 microseconds.

The primary disadvantage of present-day SCS devices is the limited current and voltage range. Typical SCSs are rated at about 50 to 300 milliamperes anode current with voltages of about 100 volts.

GTO—Gate Turn-off Switch

The GTO, a three-terminal device, is similar in construction and operation to the SCR and SCS. It is essentially a half-wave rectifier with a gate terminal for turn-on and turn-off control. The GTO is a medium-power device; the maximum anode current is about 10 amperes. However, the peak reverse working voltage ratings (anode-to-cathode and gate-to-cathode) are quite low, on the order of 16 to 20 volts.

An important characteristic of the GTO is high-speed turn-on and turn-off (about 1 microsecond). This makes the GTO a useful element for pulse-generator, multivibrator, and other related circuits.

Continued on next page.

Fig. 1-17. Symbols and descriptions for thyristor family.

LASCR—Light Activated Silicon Controlled Rectifier

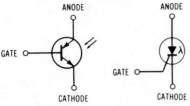

The LASCR is a unique three-terminal SCR which can be turned on by either a beam of light or a gate trigger pulse. Used primarily in optical light controls for motor and digital applications, LASCRs are low-power devices with forward current ratings of about 3 amperes.

UJT—Unijunction Transistor

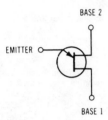

Invented about the same time as the bipolar transistor, the UJT may be considered to be a member of both the transistor and thyristor families. A three-terminal device with a single pn junction, the UJT exhibits negative resistance characteristics. Over a certain range of emitter-to-base voltage, the emitter voltage increases with decreasing emitter current.

The UJT is a highly stable switching device with operating frequencies ranging from about 1 Hz to 1 MHz. Unijunction transistors are used for a wide range of applications such as sawtooth generators, trigger circuits, timing controls, and other related circuits. Typical UJTs are low-power devices with maximum peak current and voltage ratings of about 1 ampere and 35 volts, respectively.

PUT—Programmable Unijunction Transistor

The programmable UJT (PUT) is similar to the UJT except that electrical characteristics can be "programmed" by an external circuit. The operating frequency of typical PUTs is generally in the range of 0.01 Hz to 10 kHz.

Shockley Diode

The Shockley diode is a two-terminal device that acts as an SCR without an external gate terminal. Forward conduction is switched on when the applied voltage is raised to the forward breakover voltage (V_{BO}). Shockley diodes are available with forward breakover voltage ratings of about 5 volts to 1200 volts. The maximum forward current rating (I_F) ranges from about 100 milliamperes to 1 ampere.

Shockley diodes are used primarily to trigger or switch SCRs.

SUS and SBS—Silicon Unilateral and Bilateral Switches

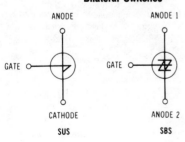

The silicon unilateral and bilateral switches, three-terminal devices, represent a new approach to the thyristor family. Designed to provide high-speed triggering of power thyristors, these devices are special integrated circuits consisting of transistors, diodes, and resistors. This type of construction provides for improved switching characteristics, stability of operation, and low fabrication costs.

The SUS is a one-way device; it can be used to switch on a forward current by two separate modes. When the applied voltage across the anode and cathode is raised above the switching voltage (V_s), forward current is immediately switched on. An input pulse to the gate terminal can also be used to turn on the SUS.

The SBS acts as a full wave rectifier; current may be switched on in either direction. Like the SUS, the SBS can be switched on by either raising the applied voltage across the anode or applying a pulse to the gate terminal.

Typical ratings for SUS and SBS devices are: switching voltage (V_s), about 8 volts or less for input to the gate terminal; forward conducting current, about 4 milliamperes; forward conducting voltage (V_F), about 1 volt.

The SUS and SBS devices are useful for many applications such as triggers for SCRs and TRIACs, digital logic, and dc latching circuits.

Chapter 2

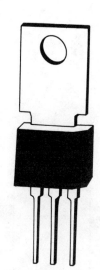

The Silicon Controlled Rectifier

THE SCR, INVENTED IN 1956, IS THE FIRST (AND probably the most important) member of the thyristor family. The SCR made possible a more reliable and efficient replacement for the older thyratron tube. In fact, the name thyristor was coined from the expression "thyratron-transistor."

The silicon controlled rectifier (SCR) is a four-layer pnpn device that possesses three individual pn junctions. Sometimes referred to as a reverse-blocking triode thyristor, the SCR is used to control or switch ac or dc power. Figure 2-1 shows the electrical symbol, basic construction, and typical packages used for SCRs. The Teccor S0301LS-2N6508 Series of SCRs, installed in TO-220AB cases, are illustrated in Fig. 2-1C. These SCRs have an on-state current rating range from 1.6 A to 25 A and maximum operating voltage range from 25 V to 600 V. The Westinghouse fast switching SCR series (T627__15) in Fig. 2-1D has ratings of 10-50 μs switching times, 150 amperes and 1200 volts maximum.

BASIC SCR OPERATION

The SCR is basically a gated silicon power rectifier. Forward load-current conduction commences when a current pulse is applied to the gate terminal. Removal of the gate current does *not* turn off forward current conduction. Thus "latched" into conduction, the SCR can only be turned off by reducing the load current below a critical value called the *holding current*.

Figure 2-2 shows how the SCR can be gated into conduction or turned off. The initial state is shown in Fig. 2-2A, the SCR is turned off and S1 is open. With no gate current, the SCR cannot conduct a load current in the forward direction. Note that the load resistor may represent a lamp, relay, heating element, or other form of a load. The function of R_g is to limit the gate current.

The conduction state is shown in Fig. 2-2B. When S1 is closed, a gate current flows. This action fires, or turns on, the SCR and allows load current to flow through the load resistor. Load current

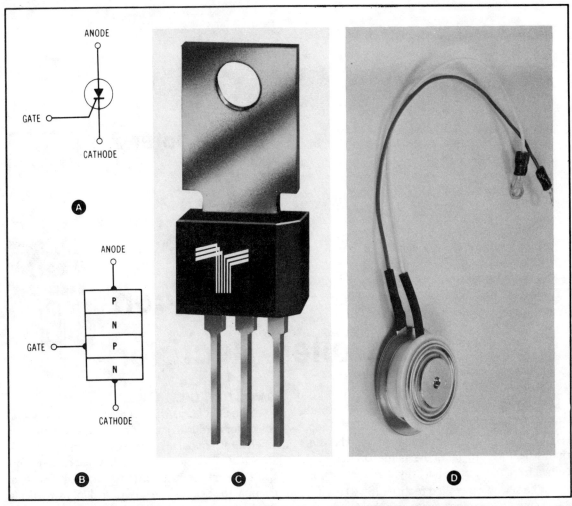

Fig. 2-1. Silicon controlled rectifier. (A) Electrical symbol. (B) Basic SCR construction. (C) TO-220 AB case (courtesy of Teccor Electronics, Inc.). (D) T62 outline (courtesy of Westinghouse Electric Corp.).

continues to flow even if S1 is opened and the gate current is turned off. Switch S1 has no further effect on load current conduction.

As shown in Fig. 2-2C, placing a shorting wire across the SCR anode and cathode terminals reduces the current though the SCR below the rated holding current (I_H). This turns off the SCR, and it will not be turned on again until S1 is closed. Note that after the SCR is turned off, the shorting wire can be removed.

An alternate method of turning off the SCR load current is shown in Fig. 2-2D. The SCR can be turned off by simply opening the anode load current. In this case, opening switch S2 will reduce the anode load current to zero. To restore load current, both S1 and S2 must be closed. You will note that S2 may be either a manual switch or relay contact.

In normal operation, the SCR does not conduct in the reverse direction. Thus, we can consider the SCR as a controlled rectifier. An important advantage of the SCR is that a very small gate current will control the switching of a large load current. For example, the 2N1597 SCR requires a gate current of only about 10 mA to switch a load current

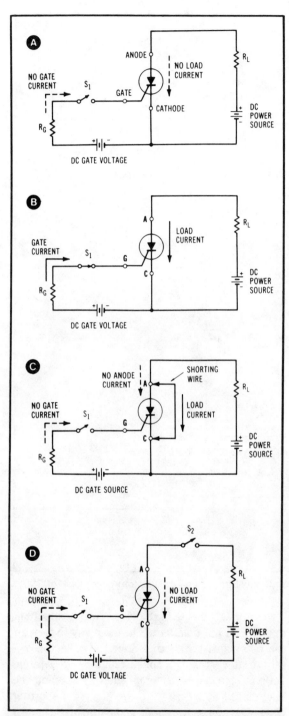

Fig. 2-2. SCR operation in dc circuits. (A) Initial state. (B) Conduction state. (C) Gate turn-off. (D) Alternate turn-off method.

of 1.6 A. In terms of power switching levels, the ratio of load power to gate switching power may be as high as 1 million to 1. Furthermore, the SCR is a fast switching device. A short pulse of gate current is adequate to turn on the forward load current of the SCR. Typical turn-on time for most SCRs is on the order of 1 to 2 μs.

The turn-off time for an SCR is more difficult to pin down. Here, you have to consider operating temperature, type of device, and the external load circuit being used. In general, the turn-off time will range from less than 10 μs to about 50 μs. Thus, an effective power switching rate for a typical SCR is about 10 to 25 kHz.

The SCR as Two Transistors

For purposes of analysis, the four-layer pnpn construction of the SCR can be visualized as two independent interconnected transistors. This is shown in Fig. 2-3A. In fact, you can build a model of the SCR with an npn and pnp transistor illustrated in Fig. 2-3B.

In order to analyze the operation within the SCR, we will connect the two-transistor model in the simple dc control circuit shown in Fig. 2-4. The values of V_G and R_G depend on the base currents required to sustain conduction.

In the initial state, S1 is open and both Q1 and Q2 are essentially cut off. Ideally, there will be no current in load resistor R_L, and the SCR will act as an open circuit between anode and cathode. In actual practice, there is a small leakage current through the two transistors.

When S1 is closed, the gate supply voltage (V_G) causes a small current (I_{B2}) to flow from the base of Q2. From transistor theory, you may recall that beta, the current gain from base to collector (I_C/I_B), may range from less than 50 to greater than 500. We can assume that the pnp and npn transistors in this circuit have a beta much greater than unity.

The gate trigger current (I_{GT}) supplies I_{B2}, which turns on I_{C2}, the collector current of Q2. At the same time, the emitter-base pn junction of Q1 is forward biased, turning Q1 on. Since Q1 collector current flows through the base of Q2, this causes

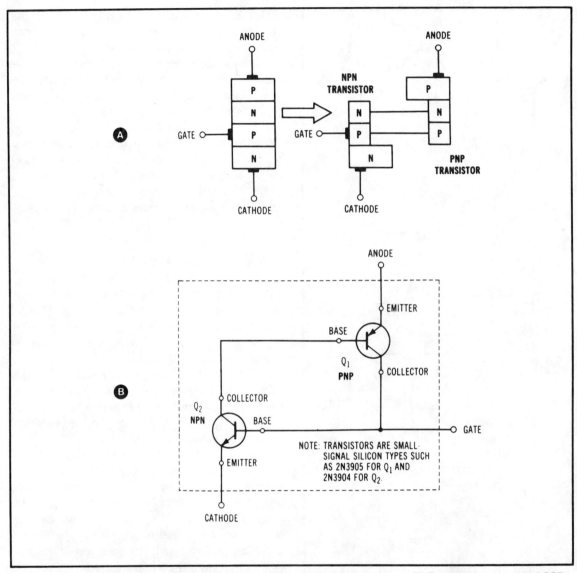

Fig. 2-3. SCR structure as two equivalent transistors. (A) Two equivalent transistors. (B) Two-transistor model of SCR.

I_{C2} to increase. The rapid increase in the collector and base currents of Q1 and Q2 is a regenerative action leading to virtual saturation of both transistors. At this time, the incoming gate trigger current is no longer needed, since the collector circuit of each transistor supplies a latching current into the base circuit of the other transistor. In this state, the SCR acts as a very low impedance, typically 0.1 ohms or less.

After Q1 and Q2 are driven into conduction, opening S1 has no effect. The only way to turn off the two transistors is to reduce the collector current to a level that will not sustain the regenerative action. This can be accomplished by opening the load circuit at switch S2 or by placing a shorting conductor across the anode and cathode terminals. This shorting circuit must reduce the SCR current below the rated *holding current* which is defined as the smallest value of current that will sustain the regenerative action. The two transistors will then go into a blocking condition. For low-power SCRs, the minimum holding current will range from less than 1 mA to about 10 mA.

In order to provide immunity from false gate signals such as noise spikes, many SCRs incorporate internal gate termination resistance. Shown as r_g in the two transistor model (Fig. 2-4), this resistance lowers the gate input impedance.

Ac Control Circuits

The SCR is an excellent switching device for ac control circuits. The ability of the SCR to be gated on during a portion of the positive half of an ac voltage waveform means that the power being applied to an ac load can be controlled. During the negative half-cycle, the SCR is turned off.

A simple SCR ac control circuit is illustrated in Fig. 2-5. The SCR is connected in the circuit to control the amount of load current (I_L) within

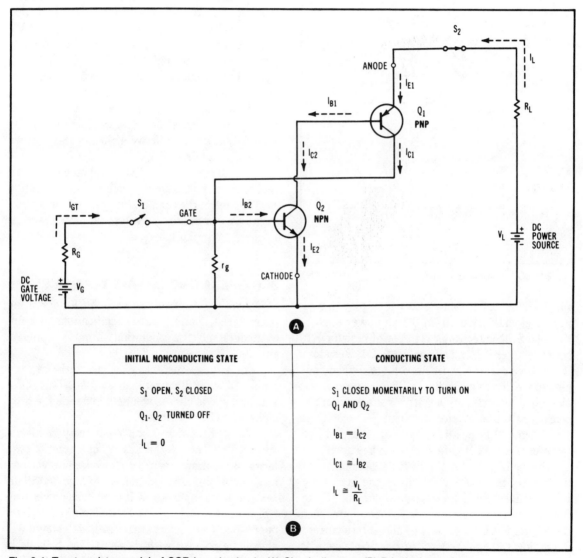

Fig. 2-4. Two-transistor model of SCR in a dc circuit. (A) Circuit diagram. (B) Principal equations.

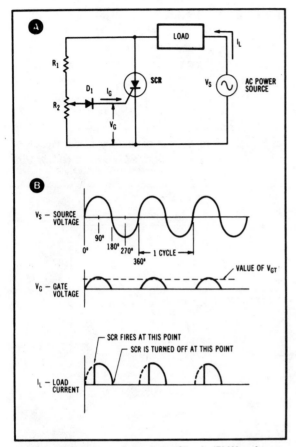

Fig. 2-5. SCR ac control circuit. (A) Circuit. (B) Waveforms.

specified limits. This simple circuit employs a gate trigger signal derived from the ac power source.

The SCR gate circuit consists of resistor R1, potentiometer R2, and silicon diode D1. Resistors R1 and R2 act as a variable voltage divider, adjustable within required limits. The purpose of the diode is to prevent any negative voltage from appearing at the SCR gate terminal.

During the positive half of the voltage waveform, the SCR gate circuit applies a varying positive voltage (V_G) to the SCR gate terminal. At the start of the positive half-cycle, the increasing gate voltage causes a rising gate current (I_G) to flow through the gate terminal. When I_g reaches the trigger level, the SCR fires and allows current (I_L) to flow through the load. At this point the SCR acts as a low impedance, and the remaining portion of the positive half-cycle is applied to the load. Although the gate circuit is effectively shorted, the absence of a gate current has no effect on the SCR conduction.

At the end of the positive half-cycle, the ac source voltage (V_S) drops to zero, and the SCR ceases to conduct. During the negative half-cycle, the voltage on the anode is negative with respect to the cathode. Thus the SCR is reverse biased and will not conduct during the negative half-cycle.

In summary, this simple ac control circuit allows load current to flow during a portion of each positive half-cycle of the ac source voltage. If R2 is adjusted for a high gate voltage, the SCR will fire early in the positive half-cycle. Thus SCR conduction will occur over almost 180 electrical degrees. On the other hand, if R2 is adjusted for a *minimum* gate voltage, the SCR turn-on will occur at approximately 90 electrical degrees. Conduction will be limited to an interval between 90 and 180 degrees for each ac cycle.

This ac control circuit is useful only in limited applications. For example, such a circuit could be used to control the temperature of a soldering iron within narrow limits. We will cover more practical ac control circuits later.

SCR OPERATING CHARACTERISTICS

Now we will investigate SCR operating characteristics and typical specifications in order to understand the operation of more complex circuits.

The voltage-current (VI) characteristics for a typical SCR are illustrated in Fig. 2-6. This set of gate and anode curves will help to define the principles of SCR operation.

Many thyristor texts and manufacturers' specifications define the SCR VI characteristics in terms of quadrant operation. This is simply a way to show positive and negative voltages and currents. For example, an SCR used in *dc* power control circuits will always be limited to *first-quadrant* operation. This means that the anode voltage will always be positive with respect to the cathode in either the conducting or nonconducting state. Note

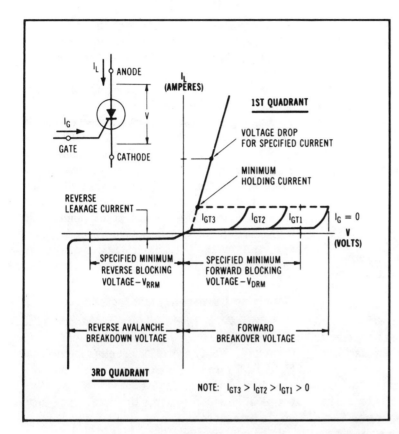

Fig. 2-6. Typical SCR VI characteristic curves.

that only forward current conduction is allowed.

When employed in ac control circuits, SCRs *conduct* current in the *first* quadrant and are *cut off* in the *third* quadrant. During some portion of the positive half-cycle, the SCR is normally turned on in first-quadrant operation. When the ac supply voltage drops to zero and passes into the negative half-cycle, the SCR is turned off. During this time period, the SCR's operating in the third quadrant. Note that in normal SCR operation, the reverse applied voltage is never allowed to exceed the minimum reverse blocking voltage. Otherwise, the SCR would break down into reverse conduction with incorrect ac control operation.

In first-quadrant operation, with no gate current (I_G equal to zero) and a moderate anode-cathode voltage, only a small leakage current flows in the SCR. However, if the anode-cathode voltage is increased beyond the maximum forward blocking voltage, the SCR will break into forward conduction. You can refer back to the two-transistor model (Fig. 2-4) to see how this happens. The two outer pn junctions are forward biased and readily conduct the leakage current; only the center pn junction is reversed biased. As the voltage across this center pn junction is increased, the electrons comprising the leakage current are accelerated to higher velocities. When the velocities of the electrons reach a critical value, any collision with nearby atoms will dislodge additional electrons. This results in an *avalanche* of current, breaking down the center pn junction. The increased current rapidly forces the SCR into full conduction. For these reasons, the SCR is almost always operated with an anode-cathode voltage less than the forward and reverse breakdown voltage ratings.

In normal operation, the SCR is turned on with a small gate current known as the gate trigger current (I_{GT}). The value of the gate trigger current varies primarily with temperature effects. Lower anode-cathode voltages require higher gate trigger currents since lower leakage currents are present in the SCR. The reverse is true; a higher anode-cathode voltage requires a lower gate trigger current. Note in Fig. 2-6 that I_{GT2} is greater than I_{GT1}.

Once fired, the SCR will remain in conduction as long as the anode, or load current is not allowed to drop below the holding current value. As stated earlier, the external circuit must reduce the anode current below the holding current rating in order to turn off the SCR.

SCR Specifications

Manufacturers of SCRs provide specifications in individual data sheets or handbooks covering the operation of their respective products. In most cases, the manufacturers have conducted extensive test and evaluation programs to insure the validity of each specification. For maximum performance, reliability, and safety, each SCR must be operated within the ratings specified by the manufacturer. Whether you are designing a new circuit or seeking a replacement SCR for an existing circuit, it is a good idea to review SCR data sheets before selecting a specific device. Most manufacturers will be glad to provide you with individual data sheets or handbooks covering their product line. Sometimes the manufacturer will offer application or engineering notes covering theory of operation, installation details, and new or unique circuits. A letter or telephone call is usually adequate to obtain this type of information.

The typical SCR data sheet lists many specifications such as maximum voltage, current, and power ratings; gate trigger-signal requirements; and operating temperature limits. These, along with the various technical expressions and symbols may be somewhat confusing at first glance. However, each individual specifications in the data sheet is important in the selection, installation, and operation of SCRs.

Some SCR specifications are closely related to those covering diodes and transistors. For example, maximum voltage, current, and power ratings fall into this category. Other specification ratings (such as gate trigger current and voltage limits, and turn-on times) are unique to SCRs and related thyristor devices. Table 2-1 lists the basic SCR operating characteristics (with the exception of thermal characteristics, which are covered in Chapter 5). Most analysis or design of SCR circuits can be accomplished using this information.

Examples of SCR specification sheets available from industry are shown in Figs. 2-7 and 2-8. These specifications cover typical medium and high-power SCRs. You will note that each set of specifications cover physical dimensions, lead configurations and thermal data in addition to operating characteristics. An interesting task at this point would be to correlate the specifications given in Figs. 2-7 and 2-8 with the basic SCR operating characteristics listed in Table 2-1.

Gating and Switching Characteristics

Figure 2-6 showed SCR turn-on characteristics in terms of gate trigger current and forward blocking voltage. Now we shall look at other factors that affect SCR turn-on. Gate trigger current and voltage levels are closely related to operating temperatures. other factors that affect gate turn-on are gate pulse triggering and the impedance of gate trigger circuits.

The relationship between gate turn-on and device operating temperature is illustrated in Fig. 2-9. This set of curves shows the operating limits for a family or type of SCR. Note that the curve OA represents the gate trigger characteristics for a specified SCR within the given family.

A basic test circuit used to obtain gate turn-on data is given in Fig. 2-10. This circuit provides for measuring the dc gate trigger current and voltage levels for the SCR at a specific temperature. With the SCR in the off-state condition, R1 is slowly reduced in value until the SCR is turned on. The voltmeter and ammeter readings indicate the values of gate voltage and current required trigger the SCR into conduction. A temperature controlled oven allows the operating temperature to be adjusted to the required levels.

One important characteristic shown in Fig. 2-9 is the maximum dc gate voltage that will *not* turn on any SCR within a given family. Sometimes re-

Table 2-1. Basic SCR Operating Characteristics.

Specification	Symbol	Description
Maximum forward or on-state anode current	I_T	The maximum allowable anode current that the SCR can withstand. Manufacturers usually define this current in terms of rms forward current (I_{RMS}), average forward current (I_{AVG}) for 180° conduction, and/or maximum surge current (I_{FM}). Surge current is specified in terms of frequency and number of cycles of current conduction. Typical values of I_{RMS} for low- and medium-power SCRs range from about 1 to 30 A. For many SCRs, momentary surge current, I_{FM}, with conduction for one-half cycle at 60 Hz is approximately 10 times the I_{RMS} current rating.
Maximum on-state or forward voltage	V_{TM}	Usually defined in terms of the SCR rated average of rms forward current, V_{TM} is the peak or maximum on-state anode-cathode voltage. Some manufacturers refer to this specification as V_F or V_{FM}. For most SCRs, V_{TM} is on the order of 1.6 V.
Maximum forward blocking voltage	V_{DRM}	The peak repetitive off-state anode-cathode voltage. Beyond the value the SCR will break down into forward conduction. This is referred to as the breakover voltage with no input gate current. For typical SCRs, V_{DRM} ranges from about 30 to 800 V.
Maximum reverse blocking voltage	V_{RRM}	The peak repetitive off-state reverse anode-cathode voltage. Beyond this value, then SCR will break down into reverse conduction. The V_{RRM} rating for typical SCRs ranges from about 30 to 800 V.
Peak forward and reverse blocking currents	I_{DRM}, I_{RRM}	The maximum forward or reverse leakage currents resulting from specified forward or reverse blocking voltages (usually V_{DRM} and V_{RRM}) and operating temperature. In some cases, the maximum allowable case or junction temperatures are specified. Typical values for I_{DRM} and I_{RRM} range from less than 1 mA for low-power SCRs to about 100 mA for high-power SCRs.
Gate trigger current	I_{GT}	The minimum forward gate current required to turn on the SCR. Since I_{GT} varies with operating temperature, load resistance, and forward blocking voltage, many SCR manufacturers specify I_{GT} in terms of these operating parameters. Low-power SCRs require an I_{GT} of about 100 to 300 μA for turn-on; the I_{GT} for medium- and high-power SCRs normally ranges from about 5 to 150 mA.
Gate trigger voltage	V_{GT}	The minimum positive gate voltage required to turn on the SCR. V_{GT} varies with operating temperature, forward blocking voltage, load resistance, and gate-cathode resistance. For temperatures of about 25°C, SCR gate trigger voltage is typically 0.7 to 0.8 V. For higher temperatures of about 100 to 125°C, V_{GT} drops to approximately 0.2 V.
Holding current	I_H	The minimum principal or load current required to maintain the SCR in the on-state. Typical values for holding current ranges from about 6 mA for low power SCRs to 80 mA for high power (65 A) SCRs.
Gate turn-on time	t_{gt}	The time required for a specified gate current to turn on the SCR. For example, one manufacturer states that for a gate-current pulse of 150 mA with a pulse width of 5 μs, its SCR will turn on in 2 μs or less. Other manufacturers rate the turn-on time in terms of a specified dc gate current, anode load current, and forward blocking voltage. Almost all SCRs possess a turn-on time in the range of 1 to 2 μs.
Turn-off time	t_q	Often referred to as commutated turn-off time; the SCR must be turned off by the external circuit. Turn-off time varies with operating temperature, anode load current, and external circuit characteristics. Most SCRs can be turned off in about 15 to 35 μs. A related characteristic, dV/dt, is often given in SCR specifications. This term defines the rate of rise of the off-state voltage with respect to time. For most SCRs, dV/dt ranges from about 50 to 300 V/μs (volts per microsecond).

ferred to as V_{GD}, this maximum nontriggering gate voltage is usually specified for the maximum, or rated, operating temperature of the device. In this example, V_{GD} is approximately 0.2 V. However, the value of V_{GD} for some SCRs may range to as low as 0.1 V. Any electrical noise spikes in the SCR gate circuit that exceed the rated V_{GD} will cause false triggering action. Thus, SCR control circuit design should take this factor into consideration.

Gate Pulse Triggering

An advantage of the SCR is that it can be turned on with a very short pulse of gate current. In most cases, this simplifies circuit design and allows for precision firing of the SCR within narrow limits. However, certain factors must be observed in using *pulsed gate triggering*.

Gate-triggering voltage (V_{GT}) and current (I_{GT}) ratings in data sheets are usually specified in terms

SILICON-CONTROLLED RECTIFIERS

... designed primarily for half-wave ac control applications, such as motor controls, heating controls and power supplies; or wherever half-wave silicon gate-controlled, solid-state devices are needed.

- Glass-Passivated Junctions
- Blocking Voltage to 600 Volts
- TO-220 Construction — Low Thermal Resistance, High Heat Dissipation and Durability

THYRISTORS

8 AMPERES RMS
200-800 VOLTS

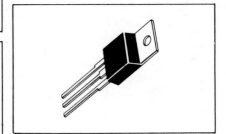

MAXIMUM RATINGS

Rating	Symbol	Value	Unit
Repetitive Peak Off-State Voltage	V_{RRM}		Volts
Repetitive Peak Reverse Voltage	V_{DRM}		
MCR218-4		200	
-5		300	
-6		400	
-7		500	
-8		600	
-9		700	
-10		800	
Forward Current RMS (All Conduction Angles)	$I_{T(RMS)}$	8.0	Amps
Peak Forward Surge Current (1/2 Cycle, Sine Wave, 60 Hz)	I_{TSM}	80	Amps
Circuit Fusing Considerations (t = 8.3 ms)	I^2t	34	A^2s
Forward Peak Gate Power	P_{GM}	5.0	Watts
Forward Average Gate Power	$P_{G(AV)}$	0.5	Watt
Forward Peak Gate Current	I_{GM}	2.0	Amps
Operating Junction Temperature Range	T_J	-40 to +125	°C
Storage Temperature Range	T_{stg}	-40 to +150	°C

THERMAL CHARACTERISTICS

Characteristic	Symbol	Max	Unit
Thermal Resistance, Junction to Case	$R_{\theta JC}$	2.0	°C/W

(1) V_{RRM} for all types can be applied on a continuous dc basis without incurring damage. Ratings apply for zero or negative gate voltage. Devices should not be tested for blocking capability in a manner such that the voltage supplied exceeds the rated blocking voltage.

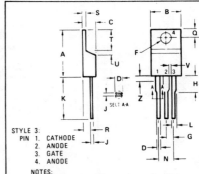

STYLE 3:
PIN 1. CATHODE
2. ANODE
3. GATE
4. ANODE

NOTES:
1. DIMENSIONS L AND H APPLIES TO ALL LEADS.
2. DIMENSION Z DEFINES A ZONE WHERE ALL BODY AND LEAD IRREGULARITIES ARE ALLOWED.
3. DIMENSIONING AND TOLERANCING PER ANSI Y14.5 1973.
4. CONTROLLING DIMENSION: INCH.

DIM	MILLIMETERS		INCHES	
	MIN	MAX	MIN	MAX
A	15.11	15.75	0.595	0.620
B	9.65	10.29	0.380	0.405
C	4.06	4.82	0.160	0.190
D	0.64	0.89	0.025	0.035
F	3.61	3.73	0.142	0.147
G	2.41	2.67	0.095	0.105
H	2.79	3.30	0.110	0.130
J	0.36	0.56	0.014	0.022
K	12.70	14.27	0.500	0.562
L	1.14	1.27	0.045	0.050
N	4.83	5.33	0.190	0.210
Q	2.54	3.04	0.100	0.120
R	2.04	2.79	0.080	0.110
S	1.14	1.39	0.045	0.055
T	5.97	6.48	0.235	0.255
U	0.76	1.27	0.030	0.050
V	1.14	-	0.045	-
Z	-	2.03	-	0.080

CASE 221A-02
TO-220 AB
All JEDEC dimensions and notes apply.

© MOTOROLA INC., 1982   DS3560

Fig. 2-7. Specifications for a typical medium-power SCR (courtesy of Motorola Semiconductor Products Inc.). (Continued through page 32.)

MCR218 series

ELECTRICAL CHARACTERISTICS (T_J = 25°C unless otherwise noted)

Characteristic	Symbol	Min	Typ	Max	Unit
Peak Forward Blocking Current (Rated V_{DRM}, T_J = 125°C)	I_{DRM}	—	—	2.0	mA
Peak Reverse Blocking Current (Rated V_{RRM}, T_J = 125°C)	I_{RRM}	—	—	2.0	mA
Peak On-State Voltage (1) (I_{TM} = 16 A Peak)	V_{TM}	—	1.5	1.8	Volts
Gate Trigger Current (Continuous dc) (V_D = 12 V, R_L = 100 Ohms)	I_{GT}	—	10	25	mA
Gate Trigger Voltage (Continuous dc) (V_D = 12 V, R_L = 100 Ohms) (Rated V_{DRM}, R_L = 1000 Ohms, T_J = 125°C)	V_{GT}	— 0.2	— —	2.5 —	Volts
Holding Current (Anode Voltage = 24 Vdc, Peak Initiating On-State Current = 0.5 A, 0.1 to 10 ms Pulse, Gate Trigger Source = 7.0 V, 20 Ohms)	I_H	—	16	30	mA
Critical Rate of Rise of Off-State Voltage (Rated V_{DRM}, Exponential Waveform, Gate Open, T_J = 125°C)	dv/dt	100	—	—	V/μs
Maximum Rate of Change of On-State Current (Rated V_{DRM}, I_{TM} = 16 A Peak, I_{GT} = 100 mA, T_J = 125°C)	di/dt	—	100	—	A/μs

(1) Pulse Test: Pulse Width = 1.0 ms, Duty Cycle ≤ 2%.

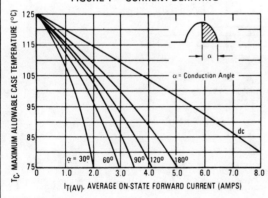

FIGURE 1 — CURRENT DERATING

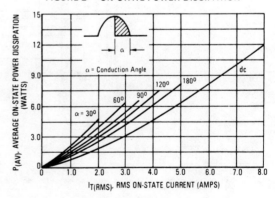

FIGURE 2 — ON-STATE POWER DISSIPATION

 MOTOROLA Semiconductor Products Inc.

MCR218 series

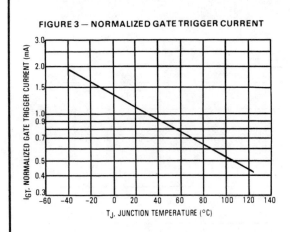

FIGURE 3 — NORMALIZED GATE TRIGGER CURRENT

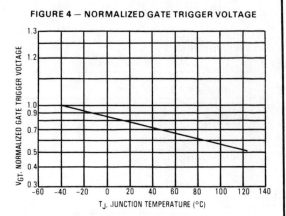

FIGURE 4 — NORMALIZED GATE TRIGGER VOLTAGE

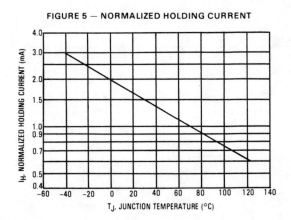

FIGURE 5 — NORMALIZED HOLDING CURRENT

Motorola reserves the right to make changes to any products herein to improve reliability, function or design. Motorola does not assume any liability arising out of the application or use of any product or circuit described herein; neither does it convey any license under its patent rights nor the rights of others.

 MOTOROLA Semiconductor Products Inc.

Fast Switching SCR T627_ _25

250 A. Avg. Up to 1200 Volts 10—50 μs

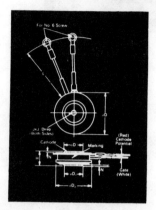

T62 Outline

Symbol	Inches		Millimeters	
	Min.	Max.	Min.	Max.
φD	1.610	1.650	40.89	41.91
φD$_1$	.745	.755	18.92	19.18
φD$_2$	1.420	1.460	36.07	37.08
H	.500	.560	12.70	14.22
φJ	.135	.145	3.43	3.68
J$_1$	.072	.082	1.83	2.08
L	7.75	8.50	196.85	215.90
N	.030		.76	

Creep distance—.34 in. (8.64 mm).
Strike distance—.52 in. min. (13.21 mm).
(In accordance with NEMA standards.)
Finish—nickel plate.
Approx. weight—2.3 oz. (66 g).
1. Dimension "H" is clamped dimension.

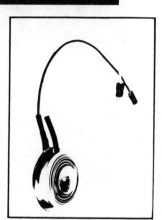

Features:
- Center fired di/namic gate
- High di/dt with soft gate control
- High frequency operation
- Sinusoidal waveform operation to 20 KHz
- Rectangular waveform operation to 20 KHz
- Low dynamic forward voltage drop
- Low switching losses at high frequency

Applications:
- Inverters for Ups Induction Heating Motor Control
- Choppers
- Crowbars

Type	Voltage	Current	Turn Off	Gate Current	Leads
T 6 2 7	1 0	2 5	6	4	D N

Ordering Information

Example
Obtain optimum device performance for your application by selecting proper Order Code.

Type T627 rated at 250 A average with V$_{DRM}$ = 1000V, I$_{GT}$ = 150 ma, tq = 20 μsec max. and flex leads—order as:

Type	Voltage		Current		Turn-off		Gate current		Leads	
Code	V$_{DRM}$ and V$_{RRM}$ (V)	Code	I$_{T(av)}$ (A)	Code	tq (unsec)	Code	I$_{GT}$ (ma)	Code	Case	Code
T627	100	01	250	25	10	8	150	4	T62	DN
	200	02			15	7				
	300	03			20	6				
	400	04			30	5				
	500	05			40	4				
	600	06			50	3				
	700	07								
	800	08								
	900	09								
	1000	10								
	1100	11								
	1200	12								

Fig. 2-8. Specifications for a typical high-power SCR (courtesy of Westinghouse Electric Corp.). (Continued through page 36.)

250 A. Avg. Up to 1200 Volts 10—50 µs

Fast Switching SCR T627_ _25

Voltage
Blocking State Maximums ① ($T_J = 125°C$)

	Symbol											
Repetitive peak forward blocking voltage, V	V_{DRM}	100	200	300	400	500	600	700	800	900	1000	1100	1200
Repetitive peak reverse voltage, V	V_{RRM}	100	200	300	400	500	600	700	800	900	1000	1100	1200
Non-repetitive transient peak reverse voltage, $t \leq 5.0$ msec, V	V_{RSM}	200	300	400	500	600	700	800	900	1000	1100	1200	1300
Forward leakage current, mA peak	I_{DRM}	← 25 →
Reverse leakage current, mA peak	I_{RRM}	← 25 →

Current
Conducting State Maximums ($T_J = 125°C$)

	Symbol	T627_ _25
RMS forward current, A	$I_{T(rms)}$	400
Ave. forward current, A	$I_{T(av)}$	250
One-half cycle surge current③, A	I_{TSM}	4500
I^2t for fusing (for times ≥ 8.3 ms) A^2 sec.	I^2t	84,000
Forward voltage drop at $I_{TM} = 625A$ and $T_J = 25°C$, V	V_{TM}	1.85
Min. repetitive di/dt④, A/µsec ①④⑥	di/dt	300

Switching
($T_J = 25°C$)

	Symbol	
Max. turn-off time, $I_T = 150A$, $T_J = 125°C$, $di_R/dt = 12.5$ $A^2/\mu sec$, reapplied dv/dt = 20V/µsec ⑦ linear to 0.8 V_{DRM}, µ sec.	t_q	10 to 50
Typ. turn-on time, $I_T = 100A$ $V_D = 100V$②, µsec.	t_{on}	3.5
Min. critical dv/dt, exponential to V_{DRM} $T_J = 125°C$, V/µsec② ①	dv/dt	300
Min. di/dt A/µsec ①④⑥	di/dt	800

Gate
Maximum Parameters ($T_J = 25°C$)

	Symbol	
Gate current to trigger at $V_D = 12V$, mA	I_{GT}	150
Gate voltage to trigger at $V_D = 12V$, V	V_{GT}	3
Non-triggering gate voltage, $T_J = 125°C$, and rated V_{DRM}, V	V_{GDM}	0.15
Peak forward gate current, A	I_{GTM}	4
Peak reverse gate voltage, V	V_{GRM}	5
Peak gate power, Watts	P_{GM}	16
Average gate power, Watts	$P_{G(av)}$	3

Thermal and Mechanical

	Symbol	
Min., Max. oper. junction temp., °C	T_J	−40 to +125
Min., Max. storage temp., °C	T_{stg}	−40 to +150
Min., Max. Mounting Force, lb.①		1000 to 1400
Max. thermal resistance, Double side cooled Junction to case, °C/Watt	$R_{\theta JC}$	.08
Case to sink, lubricated, °C/Watt	$R_{\theta CS}$	.02

① Consult recommended mounting procedures.
② Applies for zero or negative gate bias.
③ Per JEDEC RS-397, 5.2.2.1.
④ With recommended gate drive.
⑤ Higher dv/dt ratings available, consult factory.
⑥ Per JEDEC standard RS-397, 5.2.2.6.
⑦ For operation with antiparallel diode, consult factory.

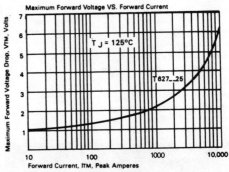

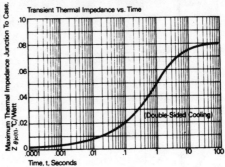

Fast Switching SCR T627_ _25

**250 A. Avg.
Up to 1200 Volts
10—50 µs**

Sinusoidal Current Data

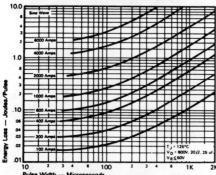

ENERGY PER PULSE FOR SINUSOIDAL PULSES

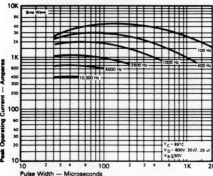

MAXIMUM ALLOWABLE PEAK ON-STATE CURRENT vs. PULSE WIDTH ($T_C = 65°C$)

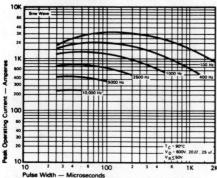

MAXIMUM ALLOWABLE PEAK ON-STATE CURRENT vs. PULSE WIDTH ($T_C = 90°C$)

Trapezoidal Wave Current Data

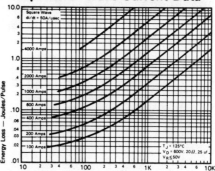

ENERGY PER PULSE FOR TRAPEZOIDAL PULSES
($di/dt = 50A/\mu sec$)

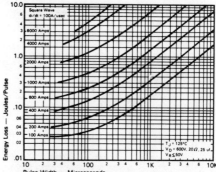

ENERGY PER PULSE FOR TRAPEZOIDAL PULSES
($di/dt = 100A/\mu sec$)

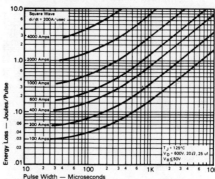

ENERGY PER PULSE FOR TRAPEZOIDAL PULSES
($di/dt = 200A/\mu sec$)

250 A. Avg. Up to 1200 Volts 10—50 µs — Fast Switching SCR T627_.25

Trapezoidal Wave Current Data ($T_C = 65°C$)

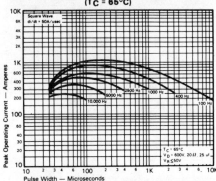

MAXIMUM ALLOWABLE PEAK ON-STATE CURRENT vs. PULSE WIDTH (di/dt = 50A/usec)

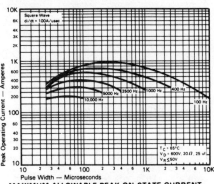

MAXIMUM ALLOWABLE PEAK ON-STATE CURRENT vs. PULSE WIDTH (di/dt = 100A/usec)

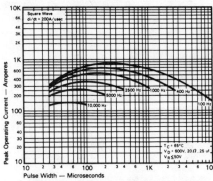

MAXIMUM ALLOWABLE PEAK ON-STATE CURRENT vs. PULSE WIDTH (di/dt = 200A/usec)

Trapezoidal Wave Current Data ($T_C = 90°C$)

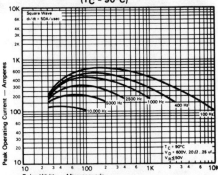

MAXIMUM ALLOWABLE PEAK ON-STATE CURRENT vs. PULSE WIDTH (di/dt = 50A/usec)

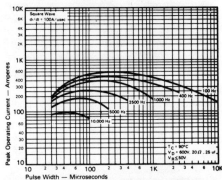

MAXIMUM ALLOWABLE PEAK ON-STATE CURRENT vs. PULSE WIDTH (di/dt = 100A/usec)

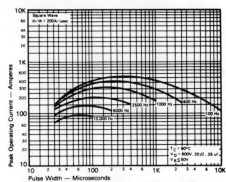

MAXIMUM ALLOWABLE PEAK ON-STATE CURRENT vs. PULSE WIDTH (di/dt = 200A/usec)

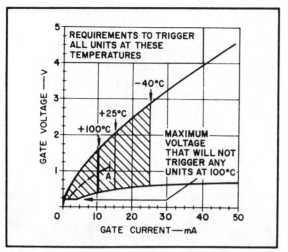

Fig. 2-9. Gate characteristic curves for a typical SCR (courtesy of RCA).

tion have a faster turn-on time—the more insensitive regions require a longer turn-on time. A higher-amplitude gate pulse causes a faster turn-on for the entire gate-cathode pn junction.

A related factor for employing a high-amplitude gate pulse signal is that of reliability. The initial conduction of current in the more sensitive regions of the gate-cathode pn junction can cause excessive localized power dissipation. This overheating effect can quickly destroy the SCR. Much research has been applied to determine the optimum geometric layout for these pn junctions and many SCR manufacturers employ special construction techniques to minimize irregular turn-on characteristics.

Ideally, the gate-trigger signal pulse width should be equal to or exceed the SCR circuit turn-on time. Since some SCR control circuits include an inductive load, the circuit turn-on time may exceed the rated turn-on time of the SCR. An analysis of the L/R time constant of the circuit will provide indication of the turn-on time. Approximately 5 L/R (five L/R time constants) is the time required to reach maximum current conduction. This design consideration, along with an optimum hard-drive gate-trigger pulse, will insure adequate SCR turn-on.

of dc values. An input gate-trigger pulse with a pulse width many times that of the SCR turn-on time is considered to be a dc level. Since the turn-on time for SCRs range from about 1 to 6 μs, most manufacturers specify V_{GT} and I_{GT} in terms of 50 μs or more. Westinghouse refers to this as a "soft drive" condition. Alternately, a gate-trigger pulse *exceeding* the specified values of trigger voltage and current is called "high" or "hard drive."

Short pulse gate-trigger signals of less than 50 μs require higher voltage and current values to achieve reliable SCR turn-on, for the following reasons. A minimum short gate-trigger pulse increases the SCR turn-on time. This is due to the fact that portions of the SCR gate-cathode pn junc-

Hard-Drive Gate-Trigger Signals

How do we determine the optimum hard-drive gate-trigger signal? Fortunately, almost all SCR data sheets list two specifications which will help us determine this gate signal: peak gate power (P_{GM}) and average gate power ($P_{G(AVG)}$) limits. These specifications relate to the repetitive firing of an SCR by a pulse train.

Let us review a few fundamentals concerning average and peak power levels contained in repetitive pulse signals. From ac circuit theory, the average power in a repetitive pulsed circuit is given by the following equation. Figure 2-11 shows the relationship among the terms given in this equation:

$$P_{AVG} = \frac{P_{PK} \times t_{on}}{t_p} \text{ watts} \quad \textbf{(Eq. 2-1)}$$

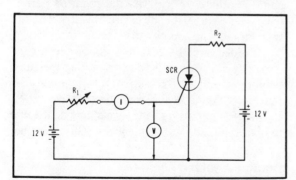

Fig. 2-10. Test circuit used to determine gate-trigger pulse requirements of thyristors (courtesy of RCA).

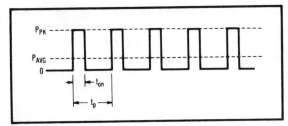

Fig. 2-11. Relationship of peak and average values for a pulse train.

where,
P_{AVG} is the average power in watts,
P_{PK} is the peak power in watts,
t_{on} is the pulse width of each pulse in seconds
t_p is the period of the pulse train seconds.

The frequency of a repetitive pulse train is expressed as:

$$f = \frac{1}{t_p} \text{ Hz} \qquad \textbf{(Eq. 2-2)}$$

where,
f is the frequency in Hertz
t_p is the period in seconds.

Given the maximum peak-power and average power gate levels for an SCR, we can solve for the gate-trigger pulse frequency in terms of gate-trigger pulse widths:

$$f = \frac{P_{G(AVG)}}{P_{GM} \times T_{on}} \text{ Hz} \qquad \textbf{(Eq. 2-3)}$$

where,
f is the pulse frequency in Hertz,
$P_{G(AVG)}$ is the average gate power in watts,
P_{GM} is the peak, or maximum allowable, gate power in watts,
t_{on} is the pulse width of each pulse in seconds.

To solve for t_{on}, given the firing frequency of the SCR circuit, simply rearrange the above terms as follows:

$$t_{on} = \frac{P_{G(AVG)}}{P_{GM} \times f} \text{ seconds} \qquad \textbf{(Eq. 2-4)}$$

Equations 2-3 and 2-4 result in providing a gate-trigger pulse more than adequate to turn on the SCR. To avoid any possibility of applying excess gate power to an SCR, it is recommended that $P_{G(AVG)}$ be derated to one-half of the specified value. One remaining caution: the data sheets will also specify peak forward gate current I_{GM} and peak forward gate voltage (V_{FGM}). Sometimes the peak reverse gate voltage (V_{RGM}) is also given. Good circuit design requires that these values never be exceeded.

GATE TURN-ON CIRCUITS

The SCR can be triggered into conduction by a variety of gate turn-on circuits. The simple dc control circuit in Fig. 2-4 employed manual switching and a dc bias supply for SCR turn-on. We noted that the ac control circuit in Fig. 2-5 uses the positive half-cycle of the ac power source to fire the SCR once each ac cycle. Although these circuits will provide positive and practical SCR turn-on, they are not adequate for most intended applications. More complex gate turn-on circuits are required for many SCR controlled power systems.

Turn-on circuits for SCRs may be divided into three basic categories—*static switching, ac phase-delay switching,* and *pulsed-gate triggering.* Each type of turn-on circuit possesses inherent advantages and limitations. In general, the required performance for each SCR control circuit will dictate the type of gate turn-on circuit to be used.

The design of the SCR gate turn-on circuit must also take into account the gate triggering specifications for the SCR. Excessive gate-triggering current or voltage levels may destroy the SCR. Insufficient gate-triggering levels may contribute to poor reliability and inadequate SCR operation.

Static Switching Circuits

Useful for either ac or dc power control, static switching circuits use either a constant or varying dc signal to turn on the SCR. In many instances, a simple resistive circuit provides the required signal level for reliable SCR turn-on. Figure 2-12 shows some basic static switching circuits that can be used for simple ac or dc control applications. In each circuit, the gate trigger signal can be derived from the power source. Each circuit can also be modified for use with independent trigger signals.

In Fig. 2-12A, S2 and S3 are normally open switches. The closure of either of these switches will fire the SCR. A fixed resistor may be used for R2 if desired. Normally-on switch S3 is used to "rearm" the circuit. Featuring high-speed switching and reliable nonmechanical operation, this type of SCR controlled dc switching is useful for industrial and other control applications.

In Fig. 2-12B, S2 and S3 are normally closed. If either switch is opened, the SCR fires and power is applied to the load. Additional sensor switches or relay contacts can be connected in series with S2 and S3 if required. Ideal for applications such as security alarm systems, this circuit draws very little current in the standby, or monitoring condition.

The circuit in Fig. 2-12C employs a bridge rectifier, D1-D4. Only one SCR is required to control both the negative and the positive half cycles of the ac power to the load. Switch S1 acts as an on-off control. Also, potentiometer R2 may be eliminated and R1 selected for proper gate current to the SCR. This circuit is useful where low-current switching for high-power loads is required.

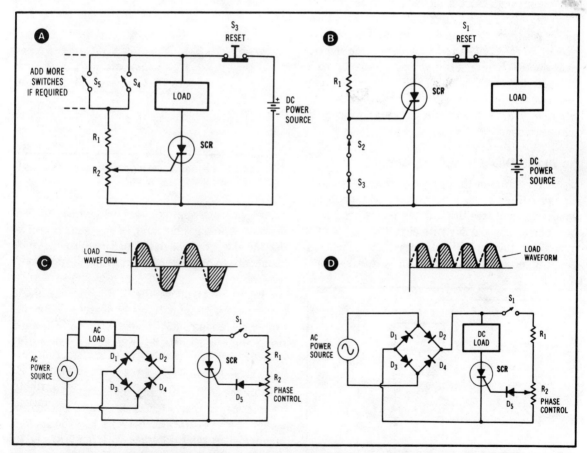

Fig. 2-12. SCR static switching circuits. (A) Normally off dc static switch. (B) Normally closed dc static switch. (C) Full-wave control of ac with one SCR. (D) Control of full-wave rectified dc with one SCR.

The circuit in Fig. 2-12D permits low-current switching of full-wave rectified power to a dc load. The phase control potentiometer, R2, allows control of the firing angle of the SCR from about 0 to 90 degrees. Alternately, R2 can be eliminated or replaced by a fixed resistance with R1 connected directly to diode D5.

A major advantage of static switching circuits is that a *small* gate current can be used to control a *large* load current. The SCRs in these circuits act as *electronic relays* without the disadvantages of arcing mechanical contacts, limited reliability, and higher cost operation. Solid-state relays using SCRs are being employed in almost all phases of electrical control applications.

The hybrid SCR-diode control circuit in Fig. 2-13 uses a high-current diode (D2) to permit full load current during negative half-cycles. Thus load power is adjustable from about one-half to full load power. In terms of electrical degrees, the hybrid SCR-diode circuit permits load current from 180-360 degrees to about 0-360 degrees. During negative half-cycles, diode D1 prevents any negative voltage from being applied to the SCR gate terminal.

Ac Phase-Delay Switching Circuits

A limitation of the simple ac static switching circuit in Fig. 2-12C is that SCR conduction for small angles of less than 90 degrees is not possible. The ac phase-delay gate circuit in Fig. 2-14A overcomes this limitation by providing a delayed gate-trigger signal. This allows an SCR firing angle

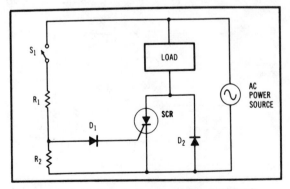

Fig. 2-13. Hybrid SCR-diode full-wave control circuit.

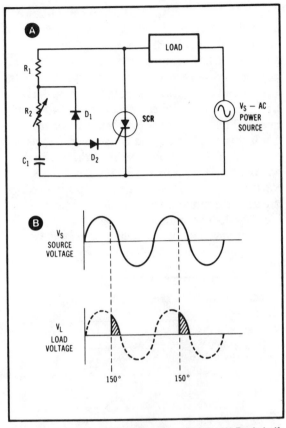

Fig. 2-14. SCR phase-shift gate trigger circuit. (A) Basic half-wave circuit. (B) Voltage waveforms.

anywhere from approximately 0 degrees to 180 degrees. The big advantage of this gate circuit is that the power applied to the load is adjustable from about zero to one-half of maximum full-wave load power.

Figure 2-14B shows the SCR gate firing angle adjusted to approximately 150 degrees. With the proper selection of R1, R2, and C1, the SCR gate firing angle is adjustable from roughly 0 to 180 degrees.

The ac lag network (R1, R2, C1) is the key to shifting the SCR gate firing angle past 90 degrees. Basic ac circuit theory states that the voltage across a capacitor in a series RC network lags the applied ac voltage. A typical lag network is illustrated in Fig. 2-15. The delayed phase angle is given by this equation:

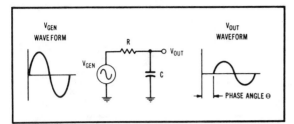

Fig. 2-15. Lag network.

$$\theta = -90 + \arctan \frac{X_C}{R} \text{ degrees} \quad \textbf{(Eq. 2-5)}$$

where,
θ is the phase angle of V_C in degrees,
X_C is the capacitive reactance of C in ohms,
R is the resistance in ohms.

When X_C equals R, the resulting phase angle is 45 degrees. This can be verified with a pocket calculator. The following example is based on a 45 degree phase shift.

EXAMPLE:
Given an operating frequency of 60 Hz and a resistance of 10,000 ohms, find the capacitance required for a phase angle of 45 degrees.

SOLUTION:

Given: R = 10,000 ohms = 10×10^3 ohms
X_C = R
Then: X_C = 10×10^3 ohms

Since

$$X_C = \frac{1}{2\pi fC} \text{ ohms} \quad \textbf{(Eq. 2-6)}$$

Then

$$C = \frac{1}{2\pi fX_C} \text{ farads}$$

Solve for C as follows:

$$C = \frac{1}{2 \times 3.1416 \times 60 \times 10 \times 10^3} F$$
$$= 2.65 \times 10^{-7} F$$
$$= 0.265 \ \mu F \cong 0.3 \ \mu F.$$

Note that the input voltage in Fig. 2-14 is adjusted to fire the SCR at approximately 150 degrees. A smaller X_C (i.e., a larger capacitor) will increase the ac phase angle closer to 90 degrees. This will result in delaying the SCR firing angle past 150 degrees.

Now let us examine a complete circuit in Fig. 2-14A. Resistors R1 and R2 comprise the total resistance (R_T) for the lag network. Adjustable resistor R2 permits a variable phase angle. During the positive half-cycle of the ac power source, D2 is forward biased and permits the delayed voltage waveform to be impressed at the gate terminal of the SCR. When this voltage reaches the required value for gate turn-on, the SCR conducts and switches current to the load. At the same time, the conducting SCR presents a low impedance to the gate trigger circuit. Accordingly, charged capacitor C1 is discharged through diode D1 and resistor R1. The small R1 C1 time constant permits C1 to be completely discharged in time for the next ac cycle.

The phase-delay gate-trigger circuit in Fig. 2-14A suffers from two limitations. Variations in the ac supply voltage will impact on the SCR firing angle, producing an undesired "jitter." Furthermore, the values of R2 and C1 may not be adequate for different ac power-supply voltage levels. For example, a phase-delay gate circuit designed for a 120 Vac power line may not be suitable for operation from a 220 or 440 Vac power line.

These problems can be eliminated or at least minimized by employing a step-down, center tap transformer in the gate-trigger circuit. As illustrated in Fig. 2-16, the secondary winding of the transformer is connected to a bridge phase-shift RC network. This circuit ensures that the gate voltage (V_G) is always one-half of the secondary voltage (V_{AB}). This is shown in the voltage vector diagram (Fig. 2-16B). Note that the voltage across the ca-

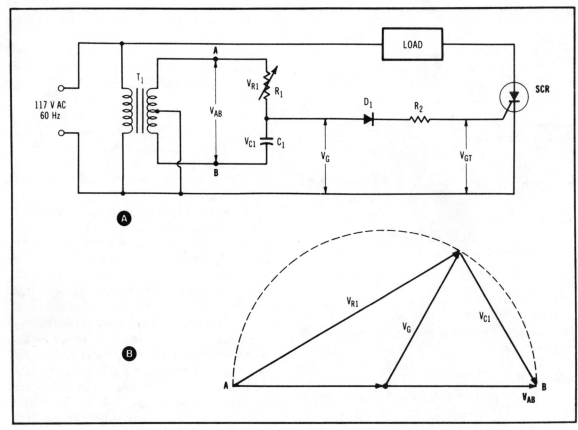

Fig. 2-16. Full 180 degree variable-phase-shift SCR ac control circuit. (A) Basic circuit for 120 Vac operation. (B) Voltage vectors.

pacitor, V_{C1}, is 90 degrees out of phase with the voltage across the resistor V_{R1}. As the value of R1 is varied from minimum to maximum, the phase angle of V_G rotates from zero to 180 degrees.

This type of circuit makes possible a standard phase-delay gate-trigger module applicable to different ac supply voltages. It is possible to use SCRs such as the 2N4444 to handle load currents of up to 8 A RMS with peak repetitive blocking voltages up to 600 V. The values of R1, R2, and C1 should be selected to provide a gate firing voltage (V_{GT}) of about 1.5 V for all desired firing angles.

Saturable Reactors as Magnetic Triggers

Sometimes referred to as *magnetic amplifiers*, *saturable reactors* have been used in ac power control systems for many years. The saturable reactor, either a single winding inductor or a multiwinding transformer, works on the principle of controlling the flow of power by varying circuit impedance. Accordingly, these devices have been applied to SCR and other thyristor trigger circuits for phase-delay firing control.

The basic SCR saturable-reactor gate circuit is illustrated in Fig. 2-17A. A low-voltage generator (V_{GEN}) supplies an alternating voltage to a series LR network. As the generator voltage increases, the buildup of magnetic flux develops an induced voltage approximately equal to the generator voltage. Accordingly, the voltage across the resistor is approximately zero. This covers the time period from T_0 to T_1 on the voltage waveforms given in Fig. 2-17B. Note that V_{GEN} has to be in phase with the ac power source. It can be derived from a step-down transformer connected to the ac power source, such as was done in Fig. 2-16A.

The saturable reactor used as the inductor in this circuit is designed so that the increasing current will eventually cause the core material to become saturated. Thus, at time T_1 further buildup of magnetic flux is prohibited by the saturated core. Therefore, the induced voltage across the inductor drops to zero, and the generator voltage appears across the resistor. This causes diode D_1 to conduct and fire the SCR.

It will be noted that a similar buildup of volt-

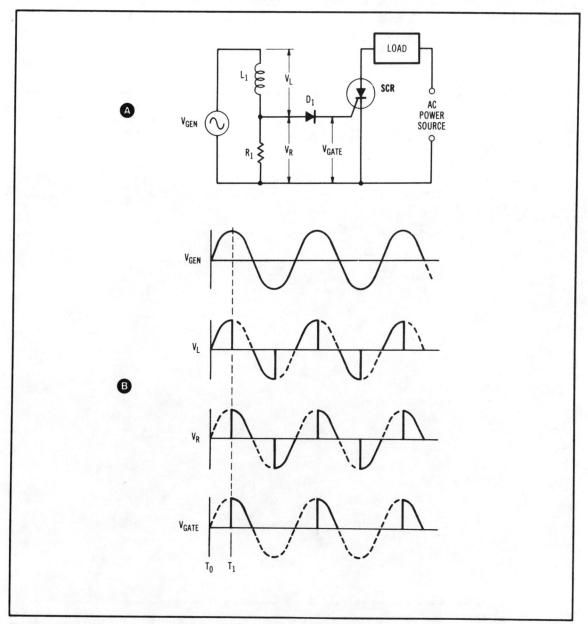

Fig. 2-17. Saturable-reactor SCR gate circuit. (A) Basic circuit. (B) Waveforms.

age occurs across the resistor during the negative half-cycle. Since D_1 is reverse biased during this time period, no negative voltage is applied to the SCR gate terminal.

The simple saturable reactor circuit in Fig. 2-17 is not very practical; it does not possess adjustable phase control characteristics. However, we can add a control winding to the saturable reactor to achieve a variable phase-delay gate-trigger voltage. Figure 2-18A shows how this concept works.

Transformer T_1 supplies a low-voltage ac signal for the SCR gate circuit. The dots by each winding on both T_1 and T_2 indicate the phase relationships between windings. For T_1, the windings are connected so that the positive half-cycle of the ac supply voltage is being applied simultaneously to both the anode terminal of the SCR and the gate trigger circuit.

The saturable-reactor transformer (T_2) is shown with two control windings and one gate winding. Commercial saturable reactors may be equipped with additional control windings. If

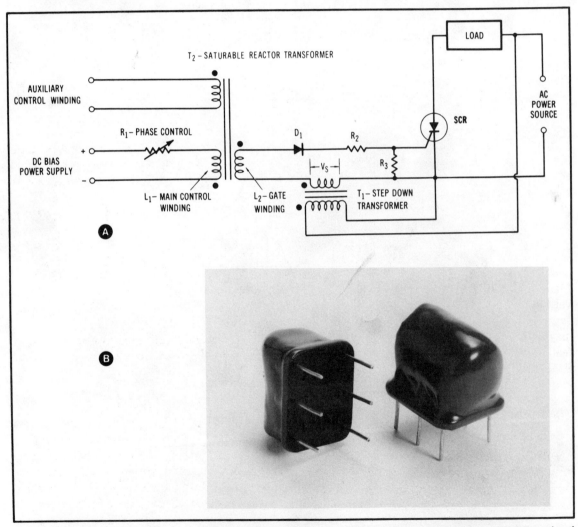

Fig. 2-18. Saturable-reactor SCR gate circuit with variable phase-delay control (courtesy of Pan-Magnetics International Inc.). (A) Diagram of typical circuit. (B) Encapsulated SCR trigger transformers.

desired, the auxiliary control windings may be used for independent control or monitoring purposes. We will use only the main control winding for our SCR control circuit.

The dc bias power supply and variable resistor provide a means for controlling the level of magnetic flux within the core material. Normally this magnetic flux is below the saturation level.

During the positive half-cycle of the ac supply voltage, diode D1 is forward biased. This permits V_s to be applied to gate winding L2 and to R2 and R3. The resulting current through L2 generates additional magnetic flux within the core material of T2, and an *opposing* induced voltage across L2. When the additional magnetic flux reaches the core saturation level, no more flux can be generated. At this time, the induced voltage drops to zero and most of V_s is applied to R2 and R3. The resulting positive voltage across R3 fires the SCR.

Voltage-divider network R2—R3 serves two purposes: the network attenuates the SCR gating signal to the proper level, and it limits the current through L2 and D1 to safe operating levels.

The SCR phase-delay control circuit in Fig. 2-18A is presented primarily to illustrate the basic principles of a half-wave self-saturating magnetic trigger circuits. Any specific circuit design must take into account the technical specifications of the particular saturable reactor and SCR being used. Many variations of this basic circuit have been developed to date. Figure 2-18B illustrates typical saturable reactor transformers available for magnetic trigger circuits. These transformers feature high-power source isolation (up to 2600 V RMS) and two or more separate windings, depending on the requirement.

Zero-Voltage Switching

Some ac SCR gate circuits employ *zero-voltage* switching in contrast to phase-delay switching. This means that the SCR is fired only at the *beginning* of a positive half-cycle, and the SCR remains in conduction for almost 180 electrical degrees. Special zero-crossing detector circuits are used to determine when the ac line voltage approaches a zero value and provide and output gate pulse. The bridge rectifier (D1—D4) supplies a full-wave rectified signal at point A. This signal keeps Q1 in a saturated state when the base voltage is about 0.7 V or higher. The effect is that Q1 collector is shorted to ground. When the base voltage drops below 0.7 V, Q1 is cut off and the collector voltage rises to about 20 V. Thus a pulse (Fig. 2-19B) is generated at each voltage crossing. Special integrated circuits such as the RCA CA3059 and the General Electric GEL300 are also available for use in zero-voltage SCR and other thyristor switching circuits. A circuit diagram and a partial specification chart for the CA3059 zero-voltage switch is given in Fig. 2-20.

Zero-voltage switching for SCR and other thyristor circuits has the important advantage of producing virtually no *radio-frequency interference* (RFI). Phase-delay switching, on the other hand, can produce substantial RFI, since switching transients at or near the peak of the ac sine wave are possible. This can cause severe interference to the reception of radio waves, particularly in the am broadcast band. We will cover this problem later in Chapter 5.

Proportional power control in zero-voltage switching SCR circuits can be achieved by firing the SCR a variable number of times during an assigned time base. A convenient time base for 60 Hz power control systems is one-half second intervals. For example, firing the SCR for 30 of each 60 positive half-cycles results in reducing the applied power to the load by one-half.

Zero-voltage switching is used primarily for power control to heating loads. Conventional metering of load current for such circuits is not practical.

Other SCR Turn-On Methods

Many types of SCR gate turn-on circuits are used in industrial control applications. Although we have covered the basic SCR turn-on methods, many new types of switching devices and circuits are available for more efficient and precise switching applications. For example, unijunction transistors, Shockley diodes, and light activated semiconductors are representative of the new devices being used in SCR and related thyristor power control cir-

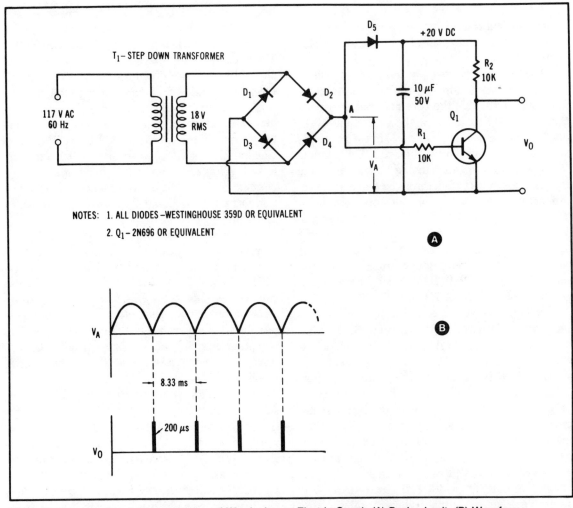

Fig. 2-19. Zero-crossing detector (courtesy of Westinghouse Electric Corp.). (A) Basic circuit. (B) Waveforms.

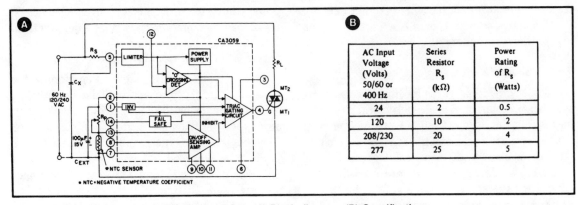

Fig. 2-20. Zero-crossing switch (courtesy of RCA). (A) Block diagram. (B) Specifications.

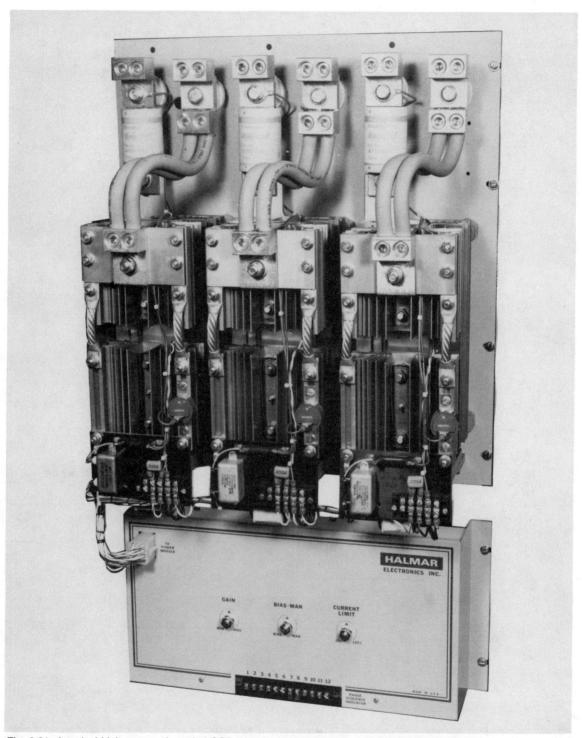

Fig. 2-21. A typical high-power, air-cooled SCR power controller (courtesy of Halmar Electronics Inc.).

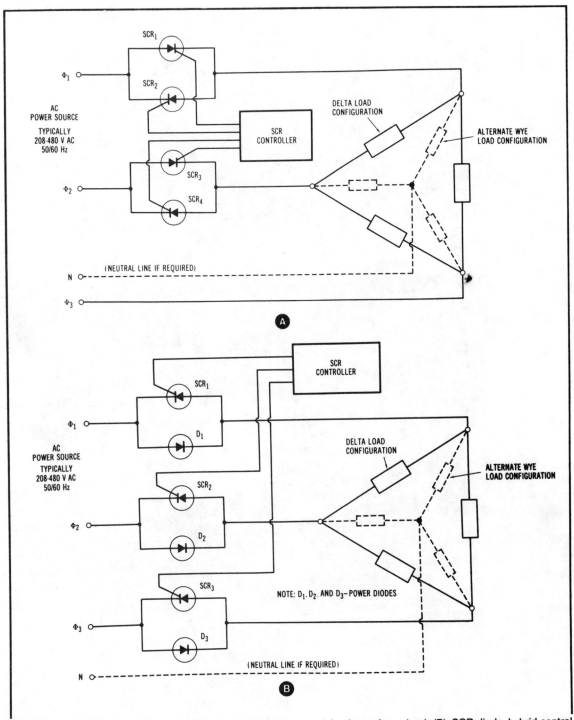

Fig. 2-22. Typical SCR controller configurations. (A) Two-leg control for three-phase load. (B) SCR-diode hybrid control for three-phase load. (C) Six-SCR control for three-phase load.

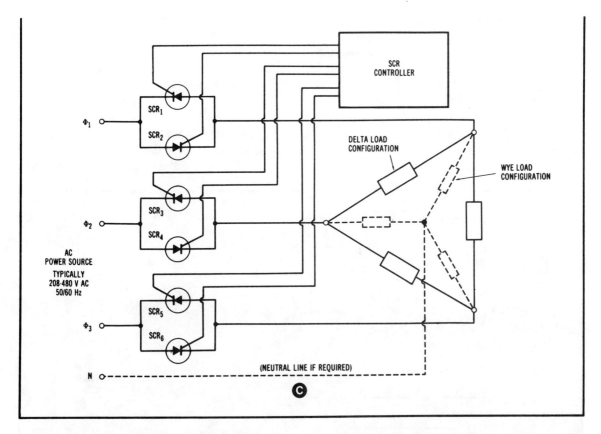

cuits. Precision firing of SCRs and other thyristors is also being accomplished by *microprocessor control*. This technique is being employed in *automated production* or "robot" machines. We will cover some of these new devices and techniques in subsequent sections of this book.

Commercial SCR Power Controllers

A wide range of SCR power controllers is available from industry. In general, these controllers are designed for operation with 120 to 480 Vac, 50 or 60 Hz, single or three-phase ac power lines. Typical load current capacities range from a few amperes to about 1000 amperes with power switching up to 600 kVA. Figure 2-21 illustrates a typical SCR controller available on the commercial market. This controller is designed for three-phase control of resistive loads with standard voltages from 208 to 600 Vac and load current up to 1000 amperes.

Commercial SCR controllers are used in many applications such as motor speed control; welding systems; heater, oven and furnace controls; battery charging systems; wire annealing manufacturing processes; and voltage regulation systems. These controllers offer many operating features including on-off control, ac or dc input control signals, potentiometer input control, dc current regulation, and soft-start operation. This last feature reduces the initial power input to the load, typically during the first half-second period. Such action protects against high inrush currents for reactive or nonlinear loads.

The high-power, three-phase SCR controllers are available in various configurations. Two-leg control of three-phase power, illustrated in Fig. 2-22A, provides an efficient and economical control technique for many three-wire delta and wye loads. Three-SCR hybrid switching systems employing power diodes as shown in Fig. 2-22B,

49

are also used for three-wire delta and wye loads. For 100-percent control of power to four-wire wye loads, a six-SCR configuration is usually specified. This circuit, shown in Fig. 2-22C, provides a symmetrical output waveform which is ideal for transformer or other inductive loads.

Three-phase power ratings of SCR power controllers for either delta or wye balanced load configurations can be computed by the following equation:

$$P_T = \sqrt{3}\, V_{LINE}\, I_{LINE} \quad \text{(Eq. 2-7)}$$

where,
P_T is the total load power in watts,
V_{LINE} is the 3-phase line voltage in volts,
I_{LINE} is the line current in amperes.

If the load characteristics are reactive, as in transformers or motors, the actual power dissipated is also a function of the power factor of the load.

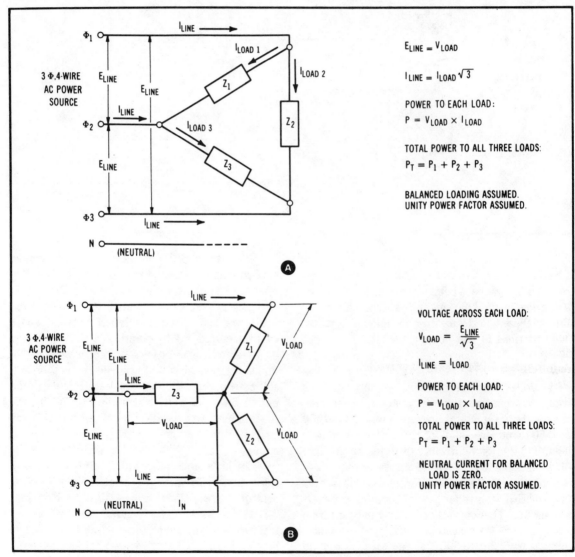

Fig. 2-23. Key relationships for three-phase delta and wye. (A) Delta load. (B) Wye load.

For resistive loads, the power factor is unity or 1. A reactive load will reduce the power factor to some value less than one. Thus P_T multiplied by the power factor represents the power dissipated by the load.

Most industrial power is furnished by three-phase four-wire, wye-connected generators. Key load equations for delta and wye-connected loads are illustrated in Fig. 2-23.

LIGHT-ACTIVATED SCRS

Light, a form of electromagnetic radiation, has been used in may solid-state applications. The *light-emitting diode* (LED) is an example of a solid-state device producing a source of light. *Photodiodes*, consisting of a single pn junction, can be used to detect the presence or absence of light. If this light-sensitive pn junction is incorporated into a transistor, the resulting configuration is a *phototransistor*. *Optoelectronics*, a term describing the marriage of the fields of optics and electronics, is beyond the scope of this book. However, we will use some of the fundamentals of optoelectronics in describing the operation of light-activated SCRs and other thyristor devices.

LASCR Characteristics

The *light activated SCR* (LASCR) is a three-terminal semiconductor device. It can be triggered into conduction by *either* a beam of light or an electrical pulse applied to the gate terminal. The electrical symbol for the LASCR and typical packaging are shown in Fig. 2-24. Typical applications for the LASCR include optical sensing, phase control, computers, and related digital electronic control systems.

Basic LASCR construction and operating characteristics are illustrated in Fig. 2-25A. The four-layer pnpn construction is similar to that of ordinary SCRs with one exception—the pn junctions are formed on a silicon pellet in an elongated manner to permit radiation by a light source. Junctions J_1 and J_3 are forward biased. Light striking these junctions has little effect, since the depletion zones are very narrow.

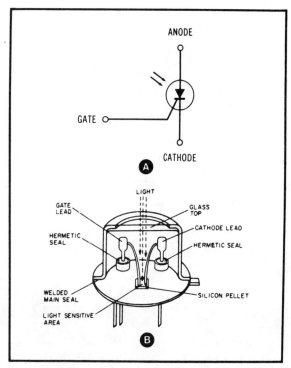

Fig. 2-24. The light activated SCR (LASCR) (courtesy General Electric Co.). (A) LASCR symbol. (B) Typical packaging.

The reverse biased junction, J_2, is the key to LASCR operation. As shown in Fig. 2-25B, this junction acts as an equivalent of a photodiode. When light strikes the wide depletion zone of J_2, absorbed photon energy produces new hole-free electron pairs. The free electrons are swept across the depletion zone toward the positive charge on the anode. This created an electrical current proportional to the light intensity. Since the remaining junctions (J_1 and J_3) are forward biased, the resulting current between the base terminals of Q_1 and Q_2 turns on both equivalent transistors. With *no* light source, the LASCR functions like a conventional SCR. The gate terminal can be used to trigger the LASCR. After the LASCR is triggered into conduction, removal of the light source and/or the electrical pulse to the gate terminal will not turn off the device. As in the case of a conventional SCR, the external circuit must be capable of turning off the LASCR.

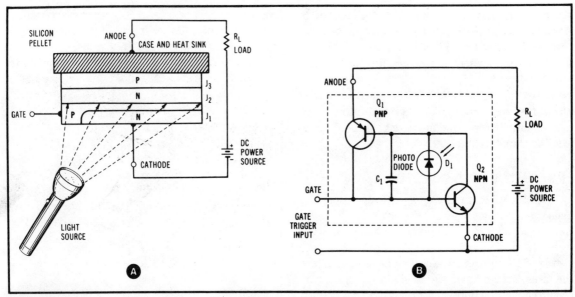

Fig. 2-25. LASCR characteristics (courtesy of General Electric Co.). (A) LASCR planar pellet. (B) LASCR equivalent circuit.

Figure 2-26A shows that the light intensity required to trigger an LASCR into conduction will vary with junction temperature. Two other variables that affect the switching characteristics are angular orientation and wavelength of the light source radiating the photosensitive junction. An

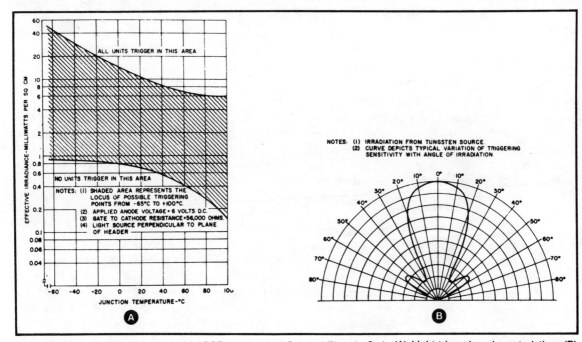

Fig. 2-26. Characteristics of a typical LASCR (courtesy of General Electric Co.). (A) Light triggering characteristics. (B) Optical angular response.

angular response pattern for a typical LASCR is given in Fig. 2-26B. This data pertains to the General Electric L8-L9 LASCR family.

The spectrum of visible light contains colors from red to violet. This corresponds to wavelengths between 0.7 micron for red down to 0.4 micron for violet. *One micron* is one-millionth of a meter, or 1×10^{-6} meter. Semiconductor photosensitive devices such as the LASCR respond more efficiently to the longer wavelengths of infrared radiation. Fortunately, most semiconductor *light emitting* devices emit light radiation with wavelengths in the infrared region.

Any design or maintenance of LASCR and other photosensitive devices must include an analysis of light source and sensor characteristics. The manufacturer's specifications for such devices will be most helpful in these actions.

Commercially available LASCRs are limited to low-power switching applications. This limitation is due primarily to the fabrication techniques required to produce light-sensitive pn junctions capable of switching the LASCR into conduction. Present day LASCRs can switch maximum rms currents of about 3 amperes. If higher-power switching is required, the LASCR can be used as a gate amplifier for conventional high-power SCRs. Figure 2-27 shows two variations of this concept. Load power in Fig. 2-27A is switched *on* when the light source is turned *on*. Note that the gate trig-

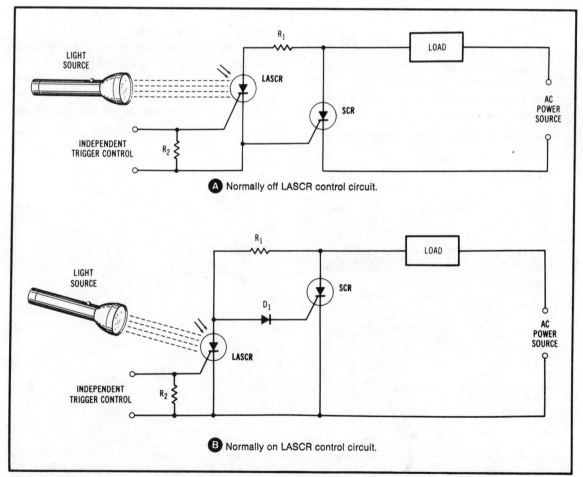

Fig. 2-27. LASCRs used as current amplifiers. (A) Normally off LASCR control circuit. (B) Normally on LASCR control circuit.

ger current to the main SCR is held to a safe operating level by resistor R1. Resistor R2 is simply a terminating resistor for the LASCR.

The main SCR in Fig. 2-27B is held in the *off* (nonconducting) state as long as the light source irradiates the LASCR. During the this time, the LASCR is turned *on* and shorts out any gate signal to the main SCR. When the light source is *removed*, the LASCR ceases to conduct during the next positive half-cycle. Hence, current through R1 will be applied to the gate terminal of the main SCR, turning it *on* and providing power to the load.

Note that the LASCR can be controlled by the independent gate-terminal connection. This is useful for emergency or backup operation.

Optoisolator SCR Circuits

Sometimes referred to as a *photocoupler*, an *optoisolator* may employ a transistor, LASCR, or other semiconductor device as the light-sensitive element. Generally, an LED is used as the light-emitting element. Figure 2-28 shows a diagram and a typical six-pin DIP configuration for an LASCR optoisolator. Since both light source and sensor are installed in the same package, no further design involving source intensity or orientation is required of the user.

A beam of light produced by the LED serves as the only coupling between the input LED and output LASCR circuit. This results in complete *electrical isolation* between input and output signals. Typically, LED trigger current is 5 to 15 mA. The LASCR forward voltage drop in the conduction state, $V_{F\,(ON)}$, is typically 1.3 V at 100 mA. The minimum holding current (I_H) is about 0.5 mA.

The electrical isolation between input and output signals of an optoisolator provides many advantages for power control systems. Isolation voltage surge ratings for these devices range up to about 3500 volts or higher. Other advantages include noise isolation, elimination of ground loop problems, simple interfacing of different input and output signal levels, and high-speed switching characteristics. These advantages, coupled with small size, high reliability, and low cost, have made

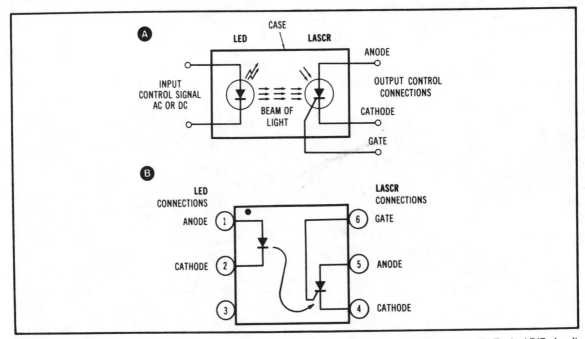

Fig. 2-28. Optoisolator employing LASCR as output control element. (A) Diagram of optoisolator. (B) Typical DIP circuit configuration.

the optoisolator a popular device in many power control systems. Figure 2-29 provides a set of specifications for a typical optoisolator or photon-coupled isolator. Figure 2-30 shows a group of typical optoisolators.

OTHER SPECIALIZED SCR DEVICES

Many specialized SCR and SCR/diode circuits are packaged for consumer and industrial use. These packages include a variety of circuits such as two SCRs in one package and combinations of SCRs and diodes for power control and rectification. Figure 2-31 shows a typical package containing two SCRs in a bridge circuit.

INTRODUCTION TO EXPERIMENTS

The experiments in this book are designed to provide basic test data on typical thyristors and to develop some practical thyristor power control circuits. You can complete these experiments with a minimum of test equipment and parts. You will need a high quality multimeter (either digital or analog), a variable dc power supply (approximately 0—15 Vdc, 1 A), and some form of experimenter breadboard to conduct the experiments. Local electronic stores (such as Heathkit, Radio Shack or others) sell test equipment, power supplies, experimenter breadboards, and most required parts. Typical equipment is shown in Figs. 2-32, 2-33, and 2-34.

Mail-order companies and parts houses are also a good source of supply for equipment and parts. A partial list of these firms is provided for your convenience:

Digikey Corporation
701 Brooks Avenue South
P.O. Box 677
Thief River Falls, MN 56701

Heathkit
Heath Company
Benton Harbor, MI 49022

JAMECO ELECTRONICS
1355 Shoreway Road
Belmont, CA 94002

Radio Shack
Division of Tandy Corporation
One Tandy Center
Fort Worth, TX 76102

A cathode ray oscilloscope is recommended in some experiments for waveform analysis. A two-channel oscilloscope, if available, is particularly useful in comparing phase delay waveforms. The oscilloscope, however, is not mandatory; you can still obtain qualitative test data using a multimeter. Also, LED indicators have been incorporated, where possible, to show presence and direction of current.

The power supply required for the experimental test circuits can be a commercial unit (such as a Heathkit IP-2718 or IP-2728) or a home built unit employing an IC variable voltage regulator. Alternately, batteries may be connected in series to perform all of the thyristor experiments.

Figure 2-35 provides a schematic for an excellent, general purpose variable dc power supply useful for many applications in addition to the thyristor experimenters. This power supply uses an LM317T IC voltage regulator (Radio Shack 276-1778 or equal) and provides an adjustable dc voltage from approximately 1.5 to 24 Vdc at 1.5 A. Also, 12.5/25 Vac, 60 Hz power is available for ac experiments involving SCRs and TRIACs. This arrangement provides a safe, isolated ac voltage source for the experimenter.

A breadboarding capability is desirable to allow easy assembly and disassembly of test circuits. One approach is to use the solderless breadboards marketed under such names as ARCHER, AP Products, HANDY, and SK Boards. Also, breadboard trainer/experimenter boxes with self-contained power supplies, function generators and LED indicators are available. The Heathkit ET-3100B is representative of these types of trainers.

In the event different power supplies are/or

SOLID STATE GENERAL ELECTRIC OPTOELECTRONICS

Photon Coupled Isolator 4N39-4N40

Ga As Infrared Emitting Diode & Light Activated SCR

The General Electric 4N39 and 4N40 consist of a gallium arsenide, infrared emitting diode coupled with a light activated silicon controlled rectifier in a dual in-line package.

absolute maximum ratings

INFRARED EMITTING DIODE		
†Power Dissipation (-55°C to 50°C)	*100	milliwatts
†Forward Current (Continuous) (-55°C to 50°C)	60	milliamps
†Forward Current (Peak) (-55°C to 50°C) (100 μsec 1% duty cycle)	1	ampere
†Reverse Voltage (-55°C to 50°C)	6	volts
*Derate 2.0mW/°C above 50°C.		

PHOTO-SCR			
†Off-State and Reverse Voltage 4N39 (-55°C to +100°C) 4N40	200 400	volts volts	
†Peak Reverse Gate Voltage (-55°C to 50°C)	6	volts	
†Direct On-State Current (-55°C to 50°C)	300	milliamps	
†Surge (non-rep) On-State Current (-55°C to 50°C)	10	amps	
†Peak Gate Current (-55°C to 50°C)	10	milliamps	
†Output Power Dissipation (-55°C to 50°C)**	400	milliwatts	
**Derate 8mW/°C above 50°C.			

TOTAL DEVICE
†Storage Temperature Range -55°C to 150°C
†Operating Temperature Range -55°C to 100°C
†Normal Temperature Range (No Derating) -55°C to 50°C
†Soldering Temperature (1/16" from case, 10 seconds) 260°C
†Total Device Dissipation (-55°C to 50°C), 450 milliwatts
†Linear Derating Factor (above 50°C), 9.0mW/°C
†Surge Isolation Voltage (Input to Output). See: Pg. 23   1500V$_{(peak)}$ 1060V$_{(RMS)}$
†Steady-State Isolation Voltage (Input to Output). See: Pg. 23   950V$_{(peak)}$ 660V$_{(RMS)}$

individual electrical characteristics (25°C) (unless otherwise specified)

INFRARED EMITTING DIODE		TYP.	MAX.	UNITS
†Forward Voltage (I_F = 10mA)	V_F	1.1	1.5	volts
†Reverse Current (V_R = 3V)	I_R	–	10	microamps
Capacitance (V = 0, f = 1MHz)		50	–	picofarads

PHOTO-SCR		MIN.	MAX.	UNITS
†Peak Off-State Voltage – V_{DM} 4N39 (R_{GK} = 10KΩ, T_A = 100°C) 4N40		200 400	– –	volts volts
†Peak Reverse Voltage – V_{RM} 4N39 (T_A = 100°C) 4N40		200 400	– –	volts volts
†On-State Voltage – V_T (I_T = 300mA)		–	1.3	volts
†Off-State Current – I_D 4N39 (V_D=200V, T_A=100°C, I_F=0, R_{GK}=10K)		–	50	microamps
†Off-State Current – I_D 4N40 (V_D=400V, T_A=100°C, I_F=0, R_{GK}=10K)		–	150	microamps
†Reverse Current – I_R 4N39 (V_R = 200V, T_A = 100°C, I_F = 0)		–	50	microamps
†Reverse Current – I_R 4N40 (V_R = 400V, T_A = 100°C, I_F = 0)		–	150	microamps
†Holding Current – I_H (V_{FX} = 50V, R_{GK} = 27KΩ)		–	200	microamps

coupled electrical characteristics (25°C)

			MIN.	MAX	UNITS
†Input Current to Trigger	V_{AK} = 50V, R_{GK} = 10KΩ	I_{FT}	–	30	milliamps
	V_{AK} = 100V, R_{GK} = 27KΩ	I_{FT}	–	14	milliamps
†Isolation Resistance (Input to Output)	V_{IO} = 500V$_{DC}$	r_{IO}	100	–	gigaohms
†Turn-On Time – V_{AK} = 50V, I_F = 30mA, R_{GK} = 10KΩ, R_L = 200Ω		t_{on}	–	50	microseconds
Coupled dv/dt, Input to Output (See Figure 13)			500	–	volts/microsec.
Input to Output Capacitance (Input to Output Voltage = 0, f = 1MHz)			–	2	picofarads

†Indicates JEDEC Registered Values.

Fig. 2-29. Specifications for a typical optoisolator or photon coupled isolator (courtesy of General Electric Co.). (Continued through page 59.)

TYPICAL CHARACTERISTICS

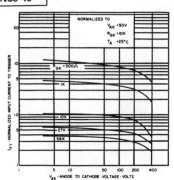

FIGURE 1. INPUT CURRENT TO TRIGGER
VS. ANODE-CATHODE VOLTAGE

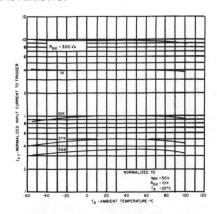

FIGURE 2. INPUT CURRENT TO TRIGGER
VS. TEMPERATURE

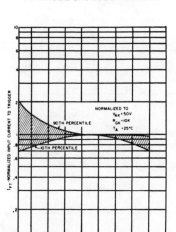

FIGURE 3. INPUT CURRENT TO TRIGGER
DISTRIBUTION VS. TEMPERATURE

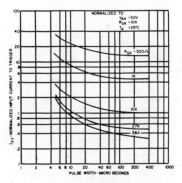

FIGURE 4. INPUT CURRENT TO TRIGGER
VS. PULSE WIDTH

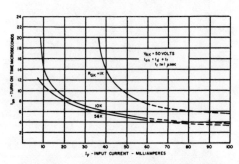

FIGURE 5. TURN-ON TIME VS. INPUT CURRENT

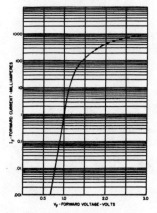

FIGURE 6. INPUT CHARACTERISTICS
I_F VS. V_F

4N39-40

TYPICAL CHARACTERISTICS

10A, T²L COMPATIBLE, SOLID STATE RELAY

Use of the 4N40 for high sensitivity, 2500V isolation capability, provides this highly reliable solid state relay design. This design is compatible with 74, 74S and 74H series T²L logic systems inputs and 220V AC loads up to 10A.

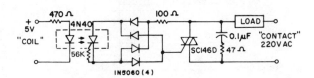

25W LOGIC INDICATOR LAMP DRIVER

The high surge capability and non-reactive input characteristics of the 4N40 allow it to directly couple, without buffers, T²L and DTL logic to indicator and alarm devices, without danger of introducing noise and logic glitches.

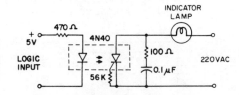

400V SYMMETRICAL TRANSISTOR COUPLER

Use of the high voltage PNP portion of the 4N40 provides a 400V transistor capable of conducting positive and negative signals with current transfer ratios of over 1%. This function is useful in remote instrumentation, high voltage power supplies and test equipment. Care should be taken not to exceed the 400 mW power dissipation rating when used at high voltages.

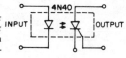

FIGURE 13
COUPLED dv/dt – TEST CIRCUIT

V_p = 800 Volts
t_p = .010 Seconds
f = 25 Hertz
T_A = 25° C

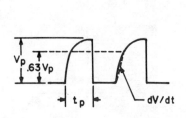
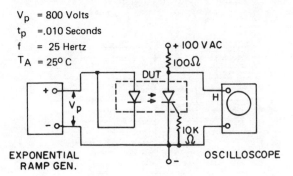

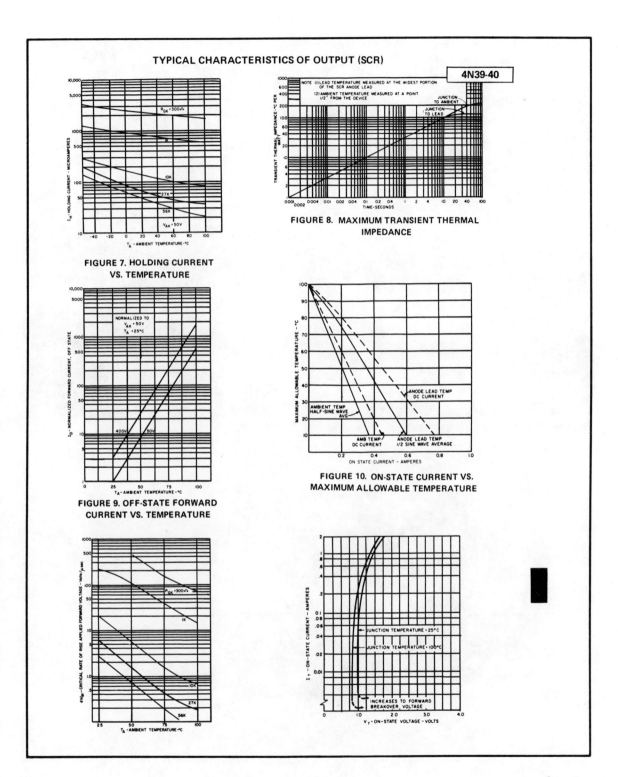

Fig. 2-30. Typical optoisolators (courtesy of Hewlett Packard).

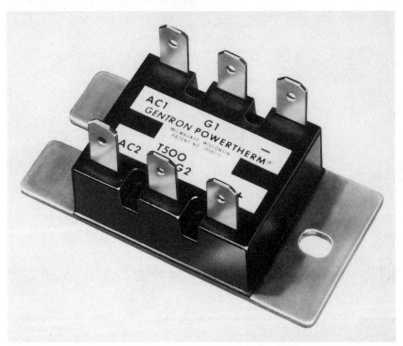

Fig. 2-31. SCR bridge circuit module (courtesy of Gentron Corp.).

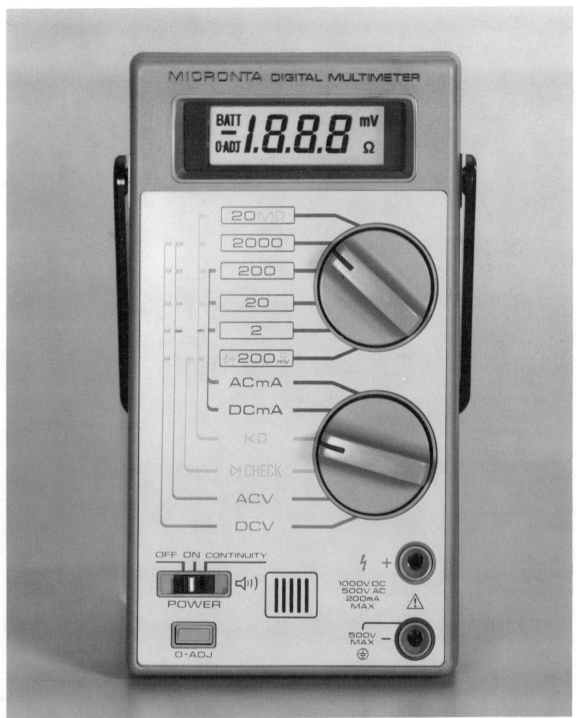

Fig. 2-32. Typical electronic test equipment (courtesy of Radio Shack, Division of Tandy Corp.) (courtesy of Heath Company). (A) Digital multimeter.

(B) One-channel cathode ray oscilloscope.

components are available to the experimenter, most of the test circuits can be modified accordingly. For example, an 18 to 24 Vdc power supply can be used if you adjust circuit component values to fit the new power levels.

CAUTION: All experiments involving 120 Vac, 60 Hz power should be completed with upmost caution. Any contact with exposed terminals or wiring connected to the 120 Vac power line can be lethal. Electrical shock can also cause burns or other injuries to the inexperienced or careless experimenter.

EXPERIMENT NO. 2-1, SCR CHARACTERISTICS

1. PURPOSE: The purpose of this experiment is to become familiar with SCR characteristics and how to use the SCR in power control circuits.

2. MATERIALS AND EQUIPMENT:

   a. Power Supply, dc Output: approximately 0—20 Vdc, 0.5—1.0 Ampere./ac Output: approximately 12/24 Vac, 0.5—1.0 Ampere.

   b. Multimeter, Analog, or Digital.

   c. SCR, 2-6 A, 200 Vdc: 2N4442, RS 276-1067 (Radio Shack) or equal.

   d. LED, Red, approximately 2 V, 20 mA: RS 276-041, NSL 5056, or equal.

   e. Potentiometer, 5 ohms, 1 Watt: RS 271-1720 or equal.

   f. Potentiometer, 10 ohms, 1 Watt: RS 271-1721 or equal.

   g. Resistors, carbon composition, two each: 330 ohms, 1/2 Watt.

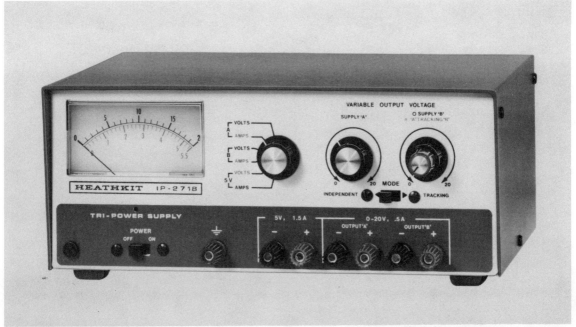

Fig. 2-33. Typical dc power supply used in electronic experimentation (courtesy of Heath Company).

h. Switch, SPST, Toggle, 3 A/125 Vac, two each: RS 275-602 or equal.

3. TEST PROCEDURES:

a. Construct the test circuit given in Fig. 2-36. You can substitute similar components in the test circuit provided you ensure proper power levels and compute test data accordingly.

b. With Switches S1 and S2 in the **off** position, adjust the power supply for 10 Vdc. Adjust Potentiometers R1 for zero ohms, close S2, R4 for zero volts between Points 1 and 2. Observe the LED indicator—it should be in the *off* (nonilluminated) state. If the LED is *on* (illuminated), check your circuit for a wiring error or a defective SCR.

c. Switch the multimeter to a low voltage dc range such as 0-5 Vdc and connect the test leads to measure the SCR gate voltage, V_{GK}. Close S1 and again confirm that the LED is still *off*. This confirms that the SCR is in a non-conducting state.

d. *Gate Trigger Voltage.* Slowly raise the SCR gate voltage (V_{GK}) by adjusting R4 until the LED switches *on*. This indicates the SCR is conducting. Observe the SCR gate voltage at the point where the SCR turns on. You may have to repeat this procedure several times to accurately obtain the value of V_{GK} which "fires" the SCR. Record this value of V_{GK} in Table 2-2 under the column V_{GT}. Remove the multimeter test leads.

e. *On-State Voltage.* With the SCR conducting (LED is *on*), measure the voltage between the SCR anode and cathode, V_{AK}. Record this value in Table 2-2.

Table 2-2. Test Data for SCR Tests.

Power Supply Volts	V_{GT} Volts	I_{GT} mA	I_{ON} mA	I_H mA	V_{AK} Volts
+10 V					
+15 V					
+20 V					

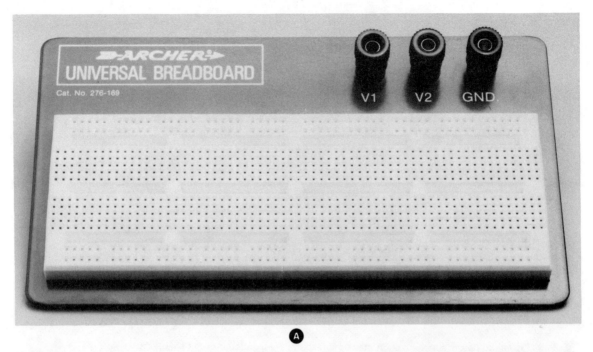

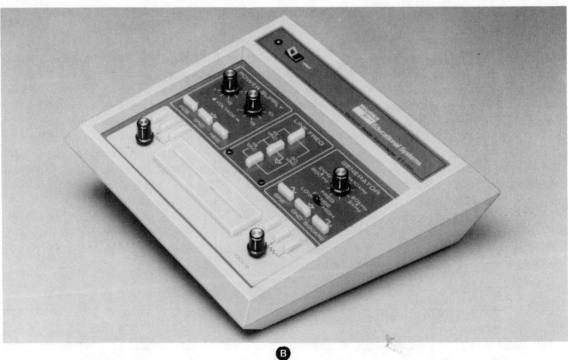

Fig. 2-34. Typical solderless breadboard and trainer for electronic experimentation (courtesy of Radio Shack, Division of Tandy Corp.) (courtesy of Heath Company). (A) Universal breadboard. (B) Electronic trainer with self-contained power supplies and signal generator.

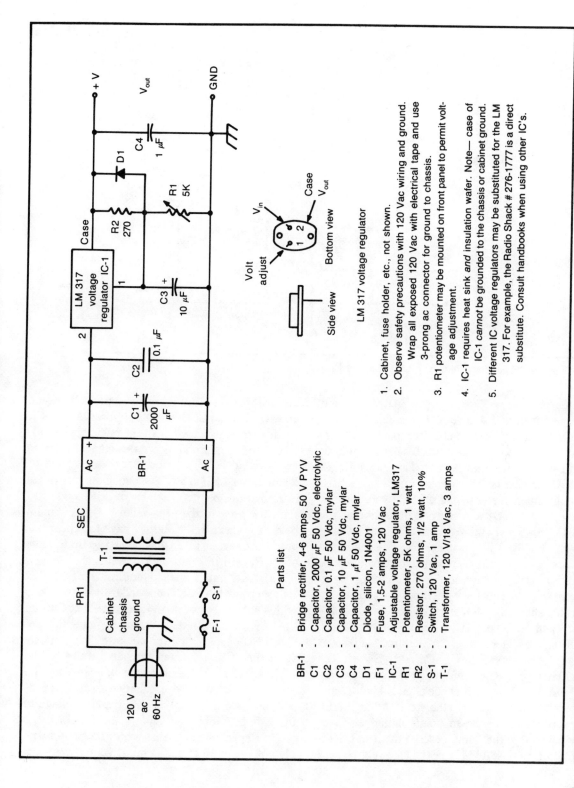

Fig. 2-35. 2-22 Vdc, 1.5 A dc power supply.

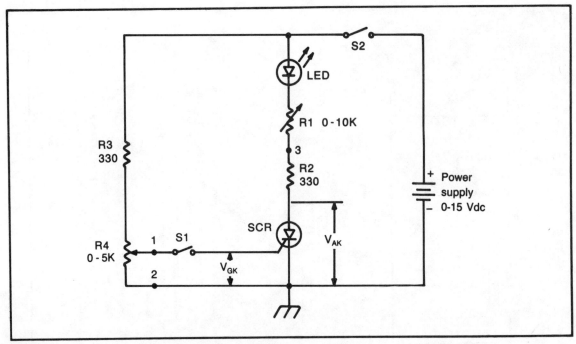

Fig. 2-36. Test circuit for determining SCR characteristics.

f. *Gate Trigger Current.* Open Switches S1 and S2, and restore R4 to zero resistance between points 1 and 2. Switch multimeter to a low dc current range such as 0-5 mA. A separate 0-5 dc milliampere meter may be used if available. Break the SCR trigger circuit at point 1 and connect the dc milliampere meter to indicate SCR gate current. Close Switches S2 and S1, and observe that the LED is *off*. Slowly increase the SCR gate voltage (V_{GK}) until the LED switches *on*. The minimum value of gate current necessary to turn on the SCR is the gate-trigger current of I_{GT}. You may have to repeat this procedure several times to obtain an accurate value of I_{GT}. Record this value in Table 2-2 and open Switches S1 and S2. Remove the dc milliampere meter and restore the circuit at point 1.

g. *Minimum Holding Current.* Switch the multimeter to a dc current range of approximately 0-5 mA. Break the SCR anode circuit at point 3 and insert the multimeter into the break. Adjust R4 for zero resistance between points 1 and 2. Close Switches S1 and S2, and observe that the LED is *off*. Slowly adjust R4 to raise the gate voltage until the LED is *on*. Return R4 to zero gate voltage and observe that the LED is still *on*. Observe the SCR anode current indicated by the dc milliampere scale on the multimeter and record this current value as I_{ON} in Table 2-2. Slowly increase the resistance of R1 until the LED is *off*. The anode current drops to zero at this point. The value of SCR anode current just *prior* to the zero value is the minimum holding current (I_H). You may have to repeat this procedure several times to obtain an accurate value of I_H. Record this data in Table 2-2.

h. An alternate method for turning off the SCR is to short out the SCR anode and cathode terminals with a short piece of wire. Readjust R1 to turn on the SCR and then reduce the gate voltage to zero. The LED should be in the *on* state. Temporarily short the SCR anode and cathode terminals. The LED should turn *off* and remain *off* even after you remove the shorting wire. In order to turn the SCR *on*, you have to reapply a gate trigger voltage to the gate terminal.

i. Repeat Steps b. through g. for power supply voltage of 15 and 20 Vdc. If your power supply does not provide these higher voltage ranges, use the highest voltage available and change Table 2-2 accordingly.

4. ANALYSIS OF TEST DATA:

The test data demonstrates the SCR turn-on and turn-off characteristics. If the SCR manufacturer's specifications are available, you can compare your test data with these specifications.

The SCR's gate trigger voltage (V_{GT}) should be 0.6-0.8 Vdc. After the SCR is turned on, the gate voltage has no further effect until the SCR is turned off. The gate trigger current should be 3-5 mA dc.

Once the SCR is turned on, the only way to turn it off is to reduce the anode or load current to a value below the minimum holding current, I_H. The SCR used in these tests has a minimum holding current or approximately 8 to 12 mA dc.

EXPERIMENT NO. 2-2, SCR AC OPERATION

1. PURPOSE: The purpose of this experiment is to investigate the operation of SCRs in three ac control circuits. These circuits cover simple, on-off half-wave operation; half-wave phase control operation; and a full-wave, SCR control circuit.

2. MATERIALS AND EQUIPMENT:

a. Power Supply, ac Output: approximately 12/24 Vac, 0.5-1.0 Ampere or transformer, 120/25 Vac, 2 A: RS 273-1512 or equal.

b. Multimeter, Analog or Digital.

c. Oscilloscope (Optional), Dual-trace, general purpose (frequency response of 5 MHz or higher). Any modern oscilloscope marketed by such companies as B&K Precision Products Group, Heath Company, Leader Instrument Company or VIZ Manufacturing will be satisfactory for these experiments.

d. Bridge rectifier, 1 A/50 Vdc: RS 272-1161 or equal.

e. Capacitor, 0.22 µF/50 Vdc: RS 272-1070 or equal.

f. Diode, Rectifier, 1 A, 200 Vdc, two each: 1N4003, RS 272-1102 or equal.

g. Fuse, 1 A/120 Vac, with In-Line Fuse Holder: RS 270-12723 and RS 2722-1281 or equal.

h. LED, Tri-Color, approximately 2 V, 25 mA: RS 276-035 or equal.

i. Potentiometer, 1 Watt, Linear Taper, 5 kohm: RS 271-1714 or equal.

j. Potentiometer, 1 Watt, Linear Taper, 1 Megohm: RS 271-211 or equal.

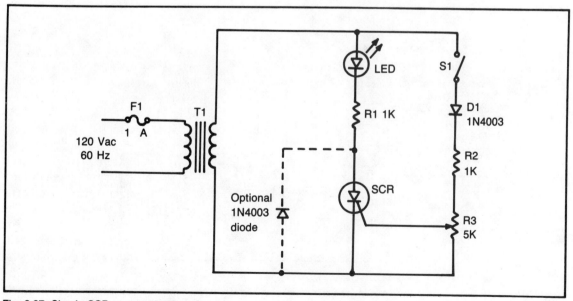

Fig. 2-37. Simple SCR ac test circuit.

k. Resistors, carbon composition, 1/2 Watt: One each—470 ohms and three each—1 kohm.

l. SCR, 2-6 A, 200 Vdc: 2N4442, RS 276-1067 or equal.

m. Switch, SPST, 3 A/125 Vac: RS 275-602 or equal.

3. TEST PROCEDURES:

a. Detailed test procedures, such as those given in Experiment 2-1 are not repeated for this and subsequent experiments. A valuable part of the learning process is developing an ability to analyze circuits and develop an adequate test procedure. Sufficient information will be provided to allow you to develop your test procedures based on the equipment and components available.

b. Analyze the test circuits and determine what test data you want to record and compare with the SCR's specifications. Record these data along with oscilloscope waveforms (if available) for your analysis.

c. Construct the test circuit given in Fig. 2-37. You can substitute similar components in the test circuit provided you ensure proper power levels and compute test data accordingly.

d. Apply power to the circuit and switch S1 to the *on* position. Adjust R3 to provide an adequate gate trigger voltage for firing the SCR. The tricolor LED should glow red, indicating the SCR is conducting current in only one direction.

e. Measure the RMS value of this gate trigger voltage. Break the SCR gate circuit and measure the RMS gate trigger current. Record these values in your test data table.

f. Optional oscilloscope experiment—observe the waveforms of V_{AK} and V_{GK} while adjusting R3 for varying firing angles. Determine the range of firing angles available with this circuit.

g. Construct the circuit given in Fig. 2-38. Apply power and confirm that the LED glows red when R3 is adjusted to firing the SCR. Vary R3 between extreme values and determine that the LED light intensity changes with a change of R3.

h. Optional oscilloscope experiment—observe the V_{AK} and V_{GK} waveforms as a function of the R3 resistance. Measure the peak voltage values of these waveforms for comparison to the SCR specifications.

i. Construct the circuit given in Fig. 2-39. If a bridge rectifier is not available, you can use four 1N4003 rectifier diodes connected as a full-wave bridge rectifier. Apply power and observe that the tricolor LED glows yellow when S1 is closed.

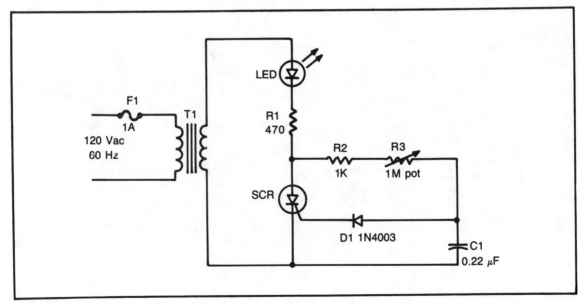

Fig. 2-38. SCR phase-control test circuit.

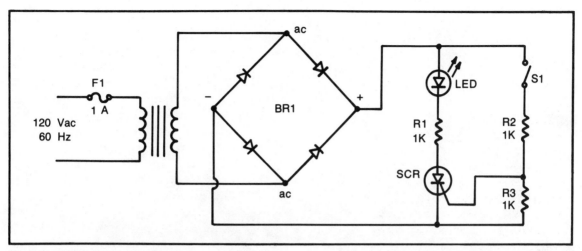

Fig. 2-39. SCR full-wave test circuit.

Measure and record the RMS voltage values of V_{AK} and V_{GK} when the SCR is conducting.

j. Optional oscilloscope experiment—measure the peak voltage values of the bridge rectifier output, V_{AK} and V_{GK}. Record these data along with waveforms of V_{AK} with S1 in the *on* and *off* positions.

Chapter 3

The Triac

THE *TRIAC* IS A THREE-TERMINAL, GATED npnpn device for controlling ac current in either direction. Originally designated as a *bidirectional triode thyristor,* it is more commonly referred to as *TRIode ac semiconductor* (TRIAC).

Either *positive or negative* gate signals may be used to trigger the triac into conduction. This characteristic helps to simplify circuit design.

Basic triac construction, the standard symbol, and typical triacs are shown in Fig. 3-1. Note that the load, or main current, terminals are designated MT_1 and MT_2. Usually, MT_1 is taken as the point of reference for voltage and current measurements made at MT_2 and the gate terminal.

Currently, triacs are available in the low to medium power-handling ranges. Maximum current and off-state voltage ratings are on the order of 40 A and 800 V, respectively.

THEORY OF OPERATION

Figure 3-2 illustrates how the n and p semiconductor sections between MT_2 and MT_2 can be visualized as equivalent parallel npnp and pnpn switches. This is similar to connecting two SCRs in parallel for bidirectional, or full-wave, current conduction. The primary difference between parallel SCRs and the equivalent switching sections of the triac lies in the gate structure and trigger methods.

Typical triac VI characteristics are given in Fig. 3-3. Operation in the first and third quadrants provides for bidirectional conduction through the device. Most manufacturers' specifications sheets contain data for each operating parameter listed in this figure.

In addition to gate triggering, the triac can be switched into conduction by two other operating conditions—exceeding the breakover voltage rating, or a sharp rise in off-state voltage. These methods of conduction are not employed in normal triac operation but they may be considered as limiting factors in circuit design. As a result, triacs switched into conduction by either of these mechanisms will not be damaged, since the triac merely

71

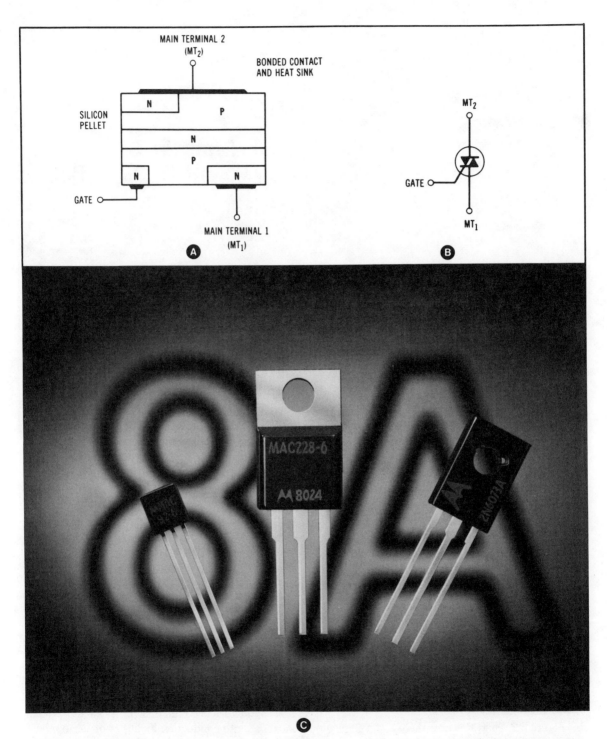

Fig. 3-1. The triac (courtesy of Motorola Semiconductor Products Inc.). (A) Basic construction. (B) Triac symbol. (C) Typical triacs.

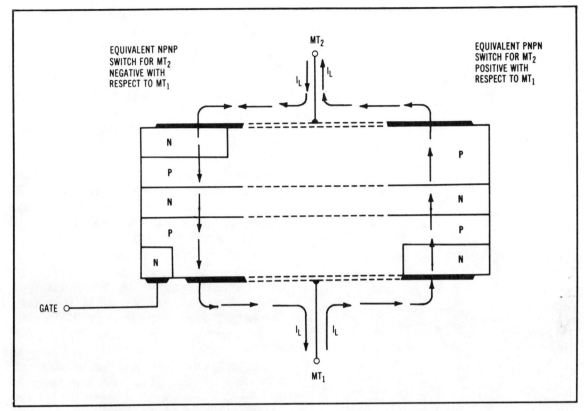

Fig. 3-2. Triac as equivalent npnp and pnpn switches.

switches to the on-state condition. However, improper circuit operation may be detrimental to the ac control system involved. Note that, in general, the triac requires no external overvoltage protection.

Voltage Breakover Turn-On

As shown on the VI characteristic curve, the triac can be switched into conduction in either the first or third quadrant by excessive voltage across the MT_2–MT_1 terminals. Triac control circuits are designed so that the rated minimum blocking voltage (V_{DRM}) is never exceeded. Manufacturers specifications normally list the V_{DRM} ratings for triacs. Transients on the ac power line can cause the off-state voltage to rise above the voltage breakover point. When this happens, leakage current through the reverse-biased pn junction avalanches, and the triac is latched into condition.

Static dV/dt Turn-On

A triac can be switched into conduction by a sharp increase of the off-state voltage. The symbol dV/dt stands for the rate of change of voltage with respect to time. It is interesting to note that the peak off-state voltage does not have to exceed the voltage breakover point for this mode of switching to occur. Generally referred to as the *critical* or *static dV/dt*, the rapid increase in voltage across the triac results in a charging current through the internal capacitances of the device. When this charging current equals or exceeds the gate trigger current (I_{GT}), the triac is triggered into conduction. The internal capacitances of a triac are illustrated in Fig. 3-4. Basic electricity theory states that the resulting current in a capacitor is a function of the capacitance and the rate of change of voltage across the capacitor. This is given by the following equation:

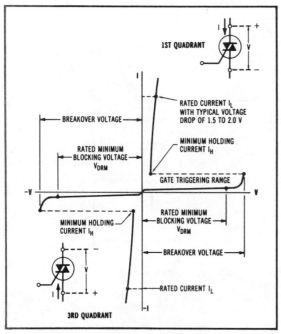

Fig. 3-3. Typical triac VI characteristic curves.

$$i = C \frac{dV}{dt} \quad \text{(Eq. 3-1)}$$

where,
i is the charging current in amperes,
C is the capacitance in farads,
dV is the voltage change in volts,
dt is the time (in seconds) associated with the voltage change.

Typical static dV/dt ratings for triacs range from about 10 V/μs for low-power devices to about 100 V/μs for higher-power devices.

A *snubber network*, consisting of a series resistor and capacitor connected across the MT_2 and MT_1 terminals, can be used to protect a triac from sharp increases in the off-state voltage. Such networks are similar to the arc-suppression networks sometimes used across relay contacts. The charging capacitor momentarily places the voltage across the resistor and the energy contained in the sharply rising portion of the voltage waveform is dissipated in the resistor. Snubber networks (Fig. 3-5) can also protect the triac against voltage transients which exceed the breakover voltage level.

Figure 3-5A illustrates how a snubber network tends to short out a transient voltage spike across a thyristor in the off-state condition. At time zero, S1 is closed and the capacitor starts to charge. Since the voltage (V_c) across the capacitor is zero at this time, C_s acts as a direct short. By the end of one time constant, V_c rises to approximately 63 percent of V_o (the supply voltage). After five time constants, V_c is approximately equal to V_o, and the charging current drops to nearly zero. The circuit in Fig. 3-5B shows how the snubber network reduces voltage transients during the off-state condition. Most triac circuits are more complex to analyze. For example, an inductive-resistive load will result in a load current lagging the applied supply voltage. Furthermore, the voltage transient pulses will "ride" on the applied ac voltage waveform. This may result in a much higher voltage peak across the triac.

The design of snubber networks must take into account peak line voltages, load characteristics, and the static dV/dt rating of the triac. In general, the time constant of the $R_s C_s$ network must be very

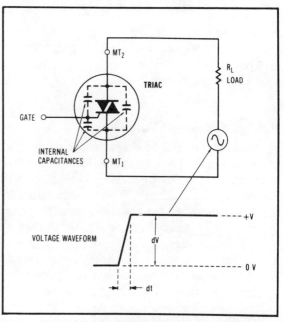

Fig. 3-4. Triac internal capacitance and static dV-dt turn-on.

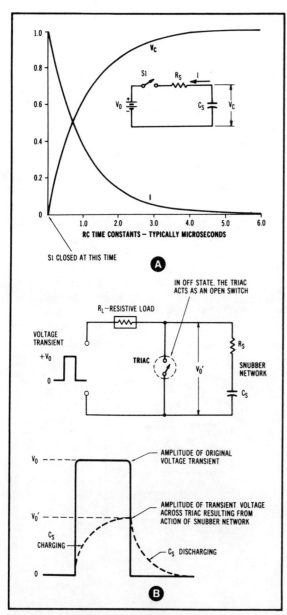

Fig. 3-5. The snubber network. (A) Charging action of capacitor in RC network. (B) Simple model of triac circuit with snubber network.

small when compared to the ac load conduction time.

TRIAC GATE CHARACTERISTICS

As started earlier, a proper positive or negative trigger signal applied to the gate terminal will switch the triac into condition. Since the triac is bidirectional, there are *four separate switching modes*—positive or negative trigger signals for conduction of either the positive or negative ac half-cycle.

In Mode I (Fig. 3-6A), the triac operates in a manner similar to that of the SCR. Junctions J_2 and J_5 are forward-biased for the ac positive half-cycle, while junction J_3 is reverse-biased. The active switching region is $P_1 N_2 P_2 N_3$. A positive trigger pulse applied to the gate terminal causes electrons to flow from N_3 into P_2. When a sufficient number of these free electrons drift across junction J_5, the triac is latched into conduction.

Mode II conduction (Fig. 3-6B), also using first quadrant operation, is triggered by a proper negative gate signal. The negative trigger signal causes a gate current to flow from N_G into P_2. Some of the free electrons will drift across the J_5 junction, increasing the leakage current through the device. This action triggers the $P_1 N_2 P_2 N_3$ switch into conduction.

Mode III operation (Fig. 3-6C) in the third quadrant is triggered into MT_1-MT_2 conduction by a negative gate signal. The reverse gate current flows through $P_2 N_G$. Free electrons emitted from N_G into P_2 increase the leakage current through the device. This causes the $P_2 N_2 P_1 N_1$ switch to latch into conduction.

Mode IV conduction (Fig. 3-6D) is similar to that of Mode III—the device operates in the third quadrant utilizing the $P_2 N_2 P_1 N_1$ switching section. Junction J_5 is reverse-biased prior to conduction. Only a small leakage current flows through the device. When the positive gate trigger signal is applied, gate current flows through N_3, P_2, and N_G. Junction J_5 is now forward-biased and permits free electrons to flow from the N_2 and P_2 sections into the N_3 section. The resulting increase in leakage current through the device latches the triac into conduction.

Modes I, II, and III exhibit the highest sensitivity characteristics; ratios of $I_{GT(I)} : I_{GT(II)} : I_{GT(III)}$ for a typical triac are on the order of 1:1.2:1.5. Mode IV has poor gate sensitivity, requiring an

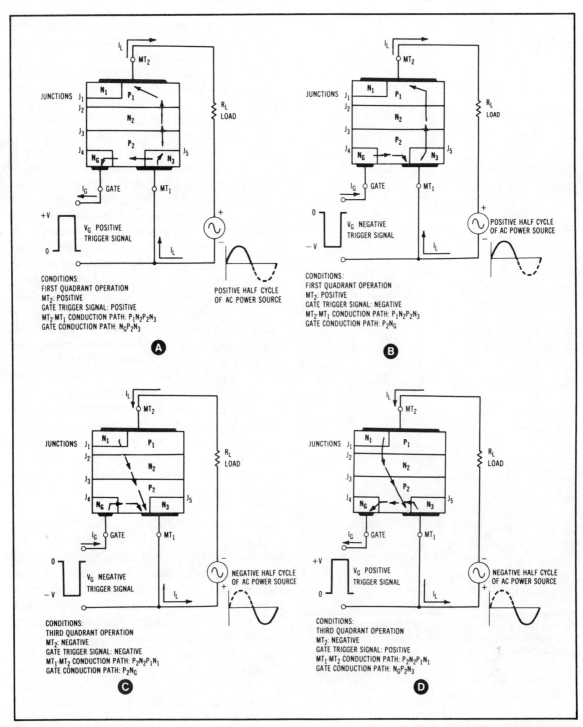

Fig. 3-6. Triac triggering modes. (A) Mode I conduction. (B) Mode II conduction. (C) Mode III conduction. (D) Mode IV conduction.

I_{GT} up to five times that of Mode I. Thus, Mode IV triggering is seldom used in triac circuit design. To illustrate mode gate-sensitivity characteristics, a typical 6 A triac requires the following I_{GT} levels for a case temperature of 25° C:

Mode: I   II   III   IV
I_{GT}:   10 mA 15 mA 20 mA 30 mA

Triac Switching Time and Commutation Considerations

Most triacs possess a gate-controlled turn-on time (t_g) on the order of 1.5 to 5 μs. Like the SCR, the triac should be triggered with a fast-rising current waveform for reliable turn-on characteristics. For example, one manufacturer recommends a gate current signal with a rise time of 1 μs, a pulse width of about 3 μs, and a peak amplitude of at least twice the rated I_{GT}. However, care should be taken to avoid exceeding the gate power-dissipation limits of the device being used.

Unlike the SCR, the triac is usually turned on *twice* each ac cycle. Thus the triac must be turned off promptly at the end of each half-cycle so that it can be turned on in the opposite direction for the next half-cycle. The successive turn-off and turn-on is referred to as *commutation*.

The switching involved with a 60-Hz ac power source may result in a commutation time of 1 ms or less. During this short interval, load current must drop below the holding current (I_H) of the triac to permit full turn-off of current. Furthermore, the triac must be gated into conduction at the proper time during the next half-cycle. With resistive loads, successive turn-off and turn-on is fairly easy to accomplish. The current in a resistive ac network is *in phase* with the applied voltage. Inductive loads, such as motors and transformers, however, pose a more difficult task for triac commutation. Figure 3-7 illustrates this problem. In the circuit shown, the snubber network is designed to prevent the triac from being turned on at points A and B where the voltage is suddenly applied between the MT_2 and MT_1 terminals. Let us look at circuit reaction to an inductive load.

The current in an inductive or inductive-resistive ac network *lags* the applied voltage. This lagging load current holds the triac in a state of conduction past the end of the ac half-cycle. When the load current drops below the holding current, the triac switches to the off-state conditions. However, by this time, the voltage associated with the next half-cycle has risen to an appreciable level. This permits a sudden increase in voltage across the triac, and this may prematurely trigger conduction during the next half-cycle. The maximum rate of rise of an off-state voltage that will not trigger the triac into conduction is known as the commutating dV/dt rating. Most triac data sheets specify the commutating dV/dt characteristic in terms of specified operating conditions such as maximum rated on-state current, maximum case temperature for the rated on-state current, and the maximum rated off-state voltage. All of these factors affect the commutating dV/dt limitation. Commutating dV/dt is usually expressed in volts per microsecond (V/μs). More critical than static dV/dt limitations, the commutating dV/dt ratings for triacs range from about 1 to 5 V/μs. Snubber networks can be used to eliminate this type of problem.

The turn-off time for the triac is seldom specified in manufacturers' data sheets as it is for the SCR. In general, the holding current for the triac and the external circuit characteristics determine the point in time of turn-off. The commutating dV/dt characteristic is a more important specification in this regard.

Operating Temperature Characteristics

As with other thyristors, the operating characteristics of the triac may vary considerably with changing temperature. All temperature-related specifications are usually referenced to case temperature. For example, gate trigger current (I_{GT}) and gate trigger voltage (V_{GT}) both vary inversely with case temperature—higher temperatures require lower-amplitude gate signals. Figure 3-8 shows the effect of varying case temperature on the gate trigger signal.

The minimum dc holding current (I_H) of the triac also varies inversely with case temperature. These characteristics are illustrated in Fig. 3-9. Dc

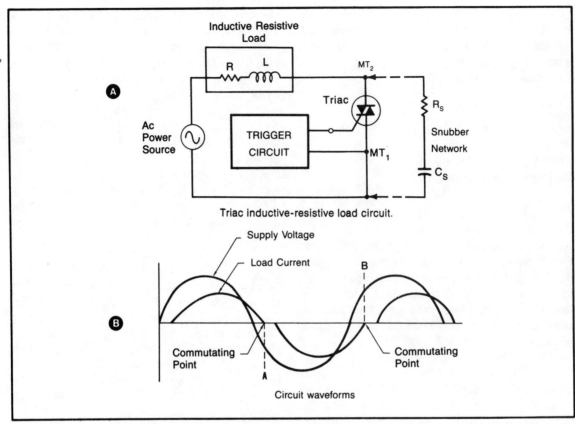

Fig. 3-7. Commutating of inductive-resistive load. (A) Triac inductive-resistance load circuit. (B) Circuit waveforms.

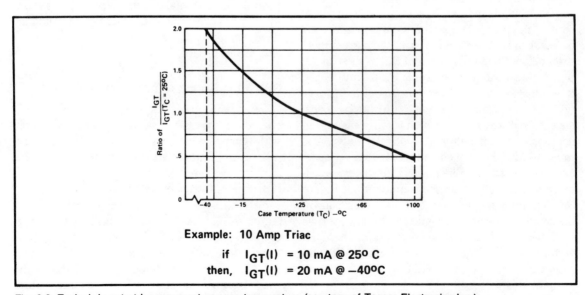

Fig. 3-8. Typical dc gate trigger current vs case temperature (courtesy of Teccor Electronics Inc.).

holding current is also related to voltage polarity across the main terminals. As a general rule, dc holding current for first quadrant operation (MT_2 positive) may exceed the third-quadrant dc holding current by 10 to 40 percent.

The design of triac control circuits requires that careful attention be given to temperature characteristics concerning such operating parameters as gate trigger signals, dc holding currents, and commutating conditions. In particular, low temperature operating environments require higher-amplitude trigger signals for reliable operation. Most manufacturers provide graphs and other data concerning the effects of temperature on device operation.

Triac Specifications

Figures 3-10 and 3-11 provide specifications for typical medium-power and high-power triacs, respectively. Like SCRs, triacs are available in a wide variety of current-handling capabilities and types of packages.

The MAC3030-40/MAC3030-40I specification contains an interesting interface circuit for use in digital control circuits. We will examine this type of interface circuit in Chapter 6.

TRIAC TURN-ON METHODS

Triacs may be triggered into conduction by a variety of methods. The particular application will generally dictate the method of triggering to be employed. As with the SCR, triac gate circuits can be designed for static, zero-voltage, or phase switching techniques. Each method offers specific advantages and limitations.

Static Switching

Triacs employed in an static switching circuits offer many advantages over mechanical switching using relays or manually operated switches. This electronic switching eliminates arcing and contact boune, both of which are problems with moving physical contacts. These factors result in more reliable operation and virtual elimination of rfi.

Basic triac static switching circuits are shown in Fig. 3-12. Numerous variations of these simple circuits have been developed for specific applications. These thyristor circuits and many more may

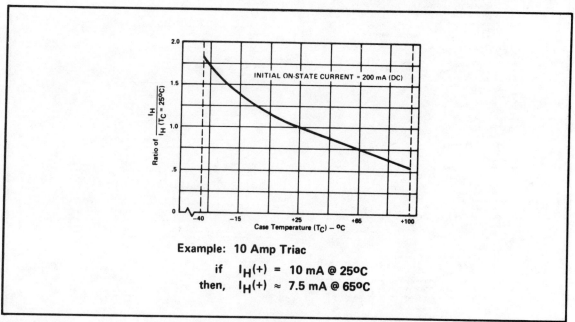

Example: 10 Amp Triac

if $I_H(+)$ = 10 mA @ 25°C

then, $I_H(+) \approx$ 7.5 mA @ 65°C

Fig. 3-9. Typical dc holding current vs case temperature (courtesy of Teccor Electronics Inc.).

TRIACS

T2800, T2801, T2802, T2850 Series

6-A and 8-A Silicon Triacs

Three-Lead Plastic Types for Power-Control and Power-Switching Applications

Features:
- 80-A and 100-A Peak Surge Full-Cycle Current Ratings
- Glass Passivated Junctions
- Short-Emitter Center-Gate Design
- Low Switching Losses
- Low Thermal Resistance
- Package Design Facilitates Mounting on a Printed-Circuit Board

Additional Features for T2850 Series:
- Internal Isolation
- Package Suitable for Direct Mounting on Heat Sink

These RCA triacs are gate-controlled full-wave silicon switches utilizing a plastic case with three leads to facilitate mounting on printed-circuit boards. They are intended for the control of ac loads in such applications as motor controls, light dimmers, heating controls, and power-switching systems.

These devices are designed to switch from an off-state to an on-state for either polarity of applied voltage with positive or negative gate triggering voltages.

The T2801 and T2802 series triacs are characterized for I^+, III^- gate triggering modes only and should suit a wide range of applications that employ diac or anode on/off triggering.

All series employ the plastic JEDEC TO-220AB package. The T2850-series package has three leads that are electrically isolated from the mounting flange. Because of this internal isolation, the triac can be mounted directly on a heat sink, without any insulating hardware; therefore heat transfer is improved and heat-sink size can be reduced.

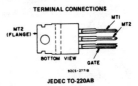

TERMINAL CONNECTIONS

JEDEC TO-220AB

MAXIMUM RATINGS, *Absolute-Maximum Values:*
For Operation with Sinusoidal-Supply Voltage at Frequencies up to 50/60 Hz and with Resistive or Inductive Load.

		T2800F T2801F T2802F T2850F	T2800A T2801A T2802A T2850A	T2800B T2801B T2802B T2850B	T2800C T2801C T2802C —	T2800D T2801D T2802D T2850D	T2800E T2801E T2802E T2850E	T2800M — T2802M —	— — T2802S —	
REPETITIVE PEAK OFF-STATE VOLTAGE:■ Gate open, $T_J = -65$ to $100°C$	V_{DROM}	50	100	200	300	400	500	600	700	V
RMS ON-STATE CURRENT (Conduction angle = 360°): Case Temperature	$I_{T(RMS)}$									
$T_C = 80°C$ (T2800, T2802, T2850 series)		←———————————— 8 ————————————→								A
= 80°C (T2801 series only)		←———————————— 6 ————————————→								A
For other conditions		←———————————— See Fig. 3 ———————————→								
PEAK SURGE (NON-REPETITIVE) ON-STATE CURRENT:	I_{TSM}									
For one cycle of applied principal voltage 60 Hz (sinusoidal), $T_C = 80°C$ (T2800, T2802, T2850 series)		←———————————— 100 ———————————→								A
50 Hz (sinusoidal) $T_C = 80°C$ (T2800, T2802, T2850 series)		←———————————— 85 ————————————→								A
60 Hz (sinusoidal), $T_C = 80°C$ (T2801 series only)		←———————————— 80 ————————————→								A
50 Hz (sinusoidal), $T_C = 80°C$ (T2801 series only)		←———————————— 65 ————————————→								A
For more than one cycle of applied principal voltage		←———————————— See Fig. 4, 5 ———————————→								
RATE OF CHANGE OF ON-STATE CURRENT: $v_D = V_{DROM}$, $I_{GT} = 200$ mA, $t_r = 0.1$ μs	di/dt	←———————————— 70 ————————————→								A/μs
FUSING CURRENT (for triac protection): At T_C shown for $I_{T(RMS)}$:										
t = 20 ms T2800, T2802, T2850		←———————————— 55 ————————————→								A^2s
T2801		←———————————— 35 ————————————→								A^2s
= 2.5 ms T2800, T2802, T2850		←———————————— 28 ————————————→								A^2s
T2801		←———————————— 18 ————————————→								A^2s
= 0.5 ms T2800, T2802, T2850		←———————————— 16 ————————————→								A^2s
T2801		←———————————— 10 ————————————→								A^2s
PEAK GATE-TRIGGER CURRENT:■ For 1 μs max. See Fig. 11	I_{GTM}	←———————————— 4 ————————————→								A
GATE POWER DISSIPATION: Peak (for 1 μs max., $I_{GTM} \leq 4$ A, See Fig. 11)	P_{GM}	←———————————— 16 ————————————→								W
AVERAGE (T2800, T2802 series)	$P_{G(AV)}$	←———————————— 0.35 ———————————→								W
AVERAGE (T2850 series)	$P_{G(AV)}$	←———————————— 0.2 ————————————→								W
TEMPERATURE RANGE: Storage	T_{stg}	←———————————— -65 to 150 ———————————→								°C
Operating (Case)	T_C	←———————————— -65 to 100 ———————————→								°C
TERMINAL TEMPERATURE (During soldering): For 10 s max. (terminals and case)	T_T	←———————————— 225 ———————————→								°C

● For either polarity of main terminal 2 voltage (V_{MT2}) with reference to main terminal 1.
■ For either polarity of gate voltage (V_G) with reference to main terminal 1.
▲ For temperature measurement reference point, see Dimensional Outline.

Fig. 3-10. Specifications for a typical low-power triac (courtesy of RCA). Continued through page 82.

TRIACS

T2800, T2801, T2802, T2850 Series

ELECTRICAL CHARACTERISTICS, At Maximum Ratings Unless Otherwise Specified, and at Indicated Temperature

CHARACTERISTICS	SYMBOL	LIMITS For All Types Except as Specified			UNITS
		MIN.	TYP.	MAX.	
Peak Off-State Current: Gate open, $T_J = 100°C$, V_{DROM} = Max. rated value	I_{DROM}	–	0.1	2	mA
Maximum On-State Voltage: (See Fig. 6, 7) For $i_T = 30$ A (peak), $T_C = 25°C$ (T2800, T2802, T2850 series) (T2801 series)	v_{TM}	– –	1.7 2	2 3	V
DC Holding Current: Gate open, Initial principal current = 150 mA (dc) $v_D = 12$ V, $T_C = 25°C$, T2800, T2850 series T2801 series T2802 series For other case temperatures	I_{HO}	– – –	15 100 20 See Fig. 8, 9, 10	30 – 60	mA
Critical Rate-of-Rise of Commutation Voltage: For $v_D = V_{DROM}$, $I_{T(RMS)} = 8$ A, commutating $di/dt = 4.3$ A/ms, gate unenergized, $T_C = 80°C$ (T2800, T2802, T2850 series) For $v_D = V_{DROM}$, $I_{T(RMS)} = 6$ A, commutating $di/dt = 4.3$ A/ms, gate unenergized, $T_C = 80°C$ (T2801 series)	dv/dt	4 2	10 10	– –	V/μs
Critical Rate-of-Rise of Off-State Voltage: For $v_D = V_{DROM}$, exponential voltage rise, gate open, $T_C = 100°C$: T2850A T2800B, T2850B T2800C, T2802C T2802D, T2850D T2800E, T2802E T2800M, T2802M T2801B T2801C T2801D T2801E	dv/dt	125 100 85 75 65 60 50 40 30 20	350 300 275 250 225 200 300 275 250 225	– – – – – – – – – –	V/μs

DC Gate-Trigger Current: For $v_D = 12$ V (dc) $R_L = 12Ω$ $T_C = 25°C$
Mode V_{MT2} V_G
I+ positive positive T2800, T2850 series
  T2801 series
  T2802 series
III− negative negative T2800, T2850 series
  T2801 series
  T2802 series
I− positive negative T2800, T2850 series only
III+ negative positive T2800, T2850 series only
For other case temperatures
	I_{GT}	–
		–
		–
		–
		–
		–
		–
		–
			10
			25
			25
			15
			25
			25
			20
			30
			See Fig. 12, 13, 14
				25
				80
				50
				25
				80
				50
				60
				60
					mA

DC Gate-Trigger Voltage: For $v_D = 12$ V (dc), $R_L = 12Ω$, $T_C = 25°C$ T2800, T2802, T2850 series T2801 series For other case temperatures For $v_D = V_{DROM}$, $R_L = 125Ω$, $T_C = 100°C$	V_{GT}	– – 0.2	1.25 1.5 See Fig. 15, 16 –	2.5 4 –	V
Gate-Controlled Turn-On Time: (Delay Time + Rise Time) For $v_D = V_{DROM}$, $I_{GT} = 80$ mA, $t_r = 0.1$ μs, $i_T = 10$ A (peak), $T_C = 25°C$ (T2800, T2802, T2850 series) (T2801 series)	t_{gt}	– –	1.6 2.2	2.5 –	μs
Thermal Resistance: Junction-to-Case (T2800, T2801, T2802 series) (T2850 series) Junction-to-Ambient	$R_{θJC}$ $R_{θJA}$	– – –	– – –	2.2 3.1 60	°C/W

● For either polarity of main terminal 2 voltage (V_{MT2}) with reference to main terminal 1.
■ For either polarity of gate voltage (V_G) with reference to main terminal 1.
▲ Variants of these devices having dv/dt characteristics selected specifically for inductive loads are available on special order; for additional information, contact your RCA Representative or your RCA Distributor.

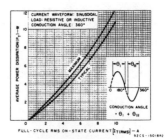

Fig. 1 — Power dissipation vs. on-state current for T2800, T2802, T2850 series.

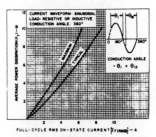

Fig. 2 — Power dissipation vs. on-state current for T2801 series.

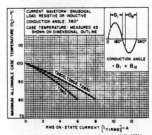

Fig. 3 — Maximum allowable case temperature vs. on-state current.

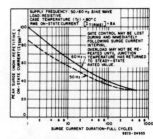

Fig. 4 — Peak surge on-state current vs. surge current duration for T2800, T2802, T2850 series.

TRIACS
T2800, T2801, T2802, T2850 Series

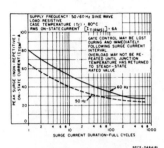

Fig. 5 — Peak surge on-state current vs. surge current duration for T2801 series.

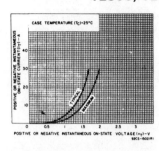

Fig. 6 — On-state current vs. on-state voltage for T2800, T2802, T2850 series.

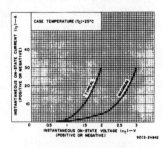

Fig. 7 — On-state current vs. on-state voltage for T2801 series.

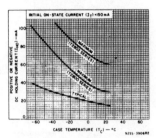

Fig. 8 — DC holding current vs. case temperature for T2800, T2802.

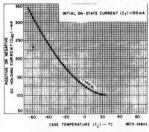

Fig. 9 — DC holding current vs. case temperature for T2801 series.

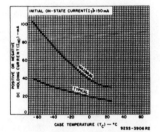

Fig. 10 — DC holding current vs. case temperature for T2850 series.

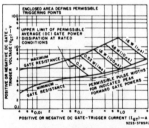

Fig. 11 — Gate pulse characteristics for all triggering modes for all series.

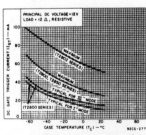

Fig. 12 — DC gate-trigger current (for I^+ and III^- triggering modes) vs. case temperature for T2800, T2802, T2850 series.

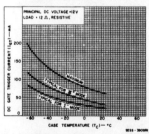

Fig. 13 — DC gate-trigger current (for I^- and III^+ triggering modes) vs. case temperature for T2800, T2802, T2850 series.

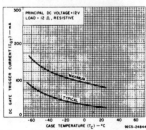

Fig. 14 — DC gate-trigger current (for I^+ and III^- triggering modes) vs. case temperature for T2801 series.

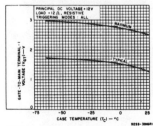

Fig. 15 — DC gate-trigger voltage vs. case temperature for T2800, T2802, T2850 series.

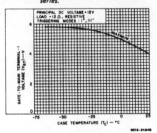

Fig. 16 — DC gate-trigger voltage vs. case temperature for T2801 series.

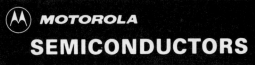

MAC3030-4

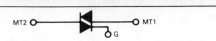

TRIACS
(THYRISTORS)

4 AMPERES RMS

SILICON BIDIRECTIONAL TRIODE THYRISTORS

...designed for full-wave, high-power control in 115 volt ac circuits, and specifically designed to be used in conjunction with the MOC3030/31 opto coupler in circuits similar to that shown below.

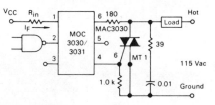

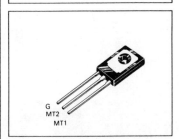

- Input to Output Isolation of 7.5 kV
- Zero Crossover Firing
- Low Drive Currents, 15 mA (MOC3031) and 30 mA (MOC3030)
- Load Can Be in Either Hot or Ground Line

MAXIMUM RATINGS

Rating	Symbol	Value	Unit
Repetitive Peak Off-State Voltage, Note 1 (T_J = –40 to +110°C) (½ Sine Wave 50 to 60 Hz, Gate Open)	V_{DRM}	250	Volts
On-State RMS Current (see Figure 1) (Full Cycle Sine Wave 50 to 60 Hz)	$I_{T(RMS)}$	4.0	Amps
Peak Nonrepetitive Surge Current (One Full Cycle, 60 Hz, T_J = 110°C)	I_{TSM}	30	Amps
Circuit Fusing Considerations (T_J = –40 to +110°C, t = 8.3 ms)	I^2t	3.6	A^2s
Peak Gate Voltage (t ≤ 2.0 μs)	V_{GM}	±5.0	Volts
Peak Gate Power (t ≤ 2.0 μs)	P_{GM}	—	Watts
Average Gate Power (T_C = 80°C, t ≤ 8.3 ms)	$P_{G(AV)}$	0.5	Watt
Peak Gate Current (t ≤ 2.0 μs)	I_{GM}	±1.0	Amp
Operating Junction Temperature Range	T_J	–40 to +110	°C
Storage Temperature Range	T_{stg}	–40 to +150	°C
Mounting Torque	—	8	in. lb.

THERMAL CHARACTERISTICS

Characteristic	Symbol	Max	Unit
Thermal Resistance, Junction to Case	$R_{\theta JC}$	3.5	°C/W
Thermal Resistance, Junction to Ambient	$R_{\theta JA}$	60	°C/W

Note 1: Ratings apply for open gate conditions. Thyristor devices shall not be tested with a constant current source for blocking voltage such that the voltage applied exceeds the rated blocking voltage.

See data sheets DS6615 and DS6616 for higher current applications.

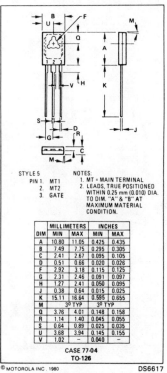

STYLE 5
PIN 1. MT1
2. MT2
3. GATE

NOTES:
1. MT = MAIN TERMINAL
2. LEADS, TRUE POSITIONED WITHIN 0.25 mm (0.010) DIA. TO DIM. "A" & "B" AT MAXIMUM MATERIAL CONDITION.

DIM	MILLIMETERS		INCHES	
	MIN	MAX	MIN	MAX
A	10.80	11.05	0.425	0.435
B	7.49	7.75	0.295	0.305
C	2.41	2.67	0.095	0.105
D	0.51	0.66	0.020	0.026
F	2.92	3.18	0.115	0.125
G	2.31	2.46	0.091	0.097
H	1.27	2.41	0.050	0.095
J	0.38	0.64	0.015	0.025
K	15.11	16.64	0.595	0.655
M	3° TYP		3° TYP	
Q	3.76	4.01	0.148	0.158
R	1.14	1.40	0.045	0.055
S	0.64	0.89	0.025	0.035
U	3.68	3.94	0.145	0.155
V	1.02	—	0.040	—

CASE 77-04
TO-126

© MOTOROLA INC. 1980

DS6617

Fig. 3-11. Specifications for a typical medium-power triac (courtesy of Motorola Semiconductor Products Inc.).

MAC3030-4

ELECTRICAL CHARACTERISTICS ($T_C = 25°C$, and Either Polarity of MT2 to MT1 Voltage unless otherwise noted)

Characteristic	Symbol	Min	Typ	Max	Unit
Peak Blocking Current (Note 1) ($V_D = 250$ V, $T_J = 110°C$)	I_{DRM}	—	—	2.0	mA
Peak On-State Voltage (Either Direction) ($I_{TM} = 6$ A Peak; Pulse Width ≤ 2.0 ms, Duty Cycle $\leq 2.0\%$)	V_{TM}	—	—	2.0	Volts
Gate Trigger Current, Continuous dc ($V_D = 12$ V, $R_L = 100$ Ω) MT2(+), G(+); MT2(−), G(−)	I_{GT}	—	—	30	mA
Gate Trigger Voltage, Continuous dc ($V_D = 12$ V, $R_L = 100$ Ω) MT2(+), G(+); MT2(−), G(−)	V_{GT}	—	—	2.0	Volts
($V_D = 250$ V, $R_L = 10$ kΩ, $T_J = 110°C$) MT2(+), G(+); MT2(−), G(−)		0.2	—	—	
Holding Current ($V_D = 12$ V, $I_{TM} = 200$ mA, Gate Open)	I_H	—	—	40	mA
Gate Controlled Turn-On Time ($V_D = 250$ V, $I_{TM} = 6$ A pk, $I_G = 100$ mA)	tgt	—	1.5	—	μs
Critical Rate of Rise of Commutation Voltage ($V_D = 250$ V, $I_{TM} = 6$ A pk, Commutating di/dt = 3.1 A/ms, Gate Unenergized, $T_C = 85°C$)	dv/dt(C)	—	5.0	—	V/μs
Critical Rate of Rise of Off-State Voltage ($V_D = 250$ V, Exponential Waveform, $T_C = 110°C$)	dv/dt	—	20	—	V/μs

Note 1: Ratings apply for open gate conditions. Thyristor devices shall not be tested with a constant current source for blocking voltage such that the voltage applied exceeds the rated blocking voltage.

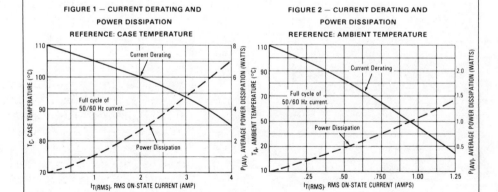

FIGURE 1 — CURRENT DERATING AND POWER DISSIPATION REFERENCE: CASE TEMPERATURE

FIGURE 2 — CURRENT DERATING AND POWER DISSIPATION REFERENCE: AMBIENT TEMPERATURE

Motorola reserves the right to make changes to any products herein to improve reliability, function or design. Motorola does not assume any liability arising out of the application or use of any product or circuit described herein; neither does it convey any license under its patent rights nor the rights of others.

 MOTOROLA Semiconductor Products Inc.

BOX 20912 • PHOENIX, ARIZONA 85036 • A SUBSIDIARY OF MOTOROLA INC.

be found in such technical magazines as *Electronics, Radio-Electronics,* and *Electronics Design.*

The static switching circuit in Fig. 3-12A illustrates the simplicity of switching a triac. Resistor R1 is used to limit the gate current and is about 100 ohms for most circuits. Switch S1 can be a manually operated on-off switch, the contacts of a small conventional or reed relay, or a transistor switch. Normally on or normally off configurations can be devised using this simple approach.

Figure 3-12B is an improved version of the simple static switch circuit. A three-position selector (S1) permits off, half-wave load-current switching, or full-wave load-current switching. A snubber network consisting of R2 and C1 is optional for inductive loads.

Remote control capability is incorporated in the static switches shown in Figs. 3-12C and 3-12D. Note that an optoisolator employing a light-activated triac is used in Fig. 3-12D. This device isolates the dc control signal from the gate circuit of the main triac.

Zero-Voltage Switching

Like SCRs, triacs can be used in zero-voltage switching circuits to control the average power applied to a load. Figure 3-13A provides a block diagram illustrating this concept. During zero-voltage switching, the triac conducts for virtually 360° of each cycle, and full power is delivered to the load. The triac is triggered at approximately the 0° and 180° points in the ac cycle.

During power-off periods, the triac is held in a nonconducting state. The ratio of power-on to power-off intervals determines the average power applied to the load. For example, the power-control time base may consist of intervals of 30 ac cycles (one-half second). If the triac is switched on for 15 full ac cycles during each one-half second interval, the average power being applied to the load is one-half of full power.

Triac zero-crossing switching circuits are used in industrial control and related applications. Like static switching, zero-crossing power switching systems are virtually free of radio-frequency-interference problems. Another important advantage is the inherent differential control capability that exists when gradual changes in average power can be applied to a load. For example, temperature control for furnaces and other heating applications is easily accomplished with zero-crossing switching and differential gate-trigger circuits.

A typical triac zero-crossing control circuit is given in Fig. 3-14. The General Electric GEL300 zero-voltage switch is employed in this circuit to trigger the triac at the 0° and 180° points of designated ac cycles.

Phase-Control Switching

Triac phase-controlled gate circuits allow conduction of load current during a specified portion of each ac half-cycle. Simple resistive gate switching circuits can be employed to trigger the triac for firing angles up to 90° in each ac half-cycle. Resistance-capacitance phase-shifting networks are used to delay the firing angle up to nearly 180°. Basic circuit configurations for these two switching methods are shown in Fig. 3-15.

These simple trigger circuits are seldom used in actual practice due to the slow-rising gate current and voltage levels. Required gate-trigger signal levels vary due to differences between triacs within a given family, the quadrant of operation, and case temperature. Such limitations would make it impossible to calibrate phase-control potentiometer R for accurate firing angles.

The performance of phase-controlled gate trigger circuits can be greatly improved by the use of a trigger device. Figure 3-16 illustrates the basic concept. For low voltage levels, the trigger device exhibits a high impedance. Except for a small leakage current, no gate signal is presented to the triac during this time. When the applied voltage is increased to the breakover level, the trigger device suddenly latches into conduction. This presents a fast-rising trigger signal to the triac, resulting in reliable turn-on of load current. Figure 4-5 (in the next chapter) shows a practical phase-controlled triac switching circuit employing a *diac* as the trigger device.

The *diac* is one of the more common trigger devices in use today. Other trigger devices used in

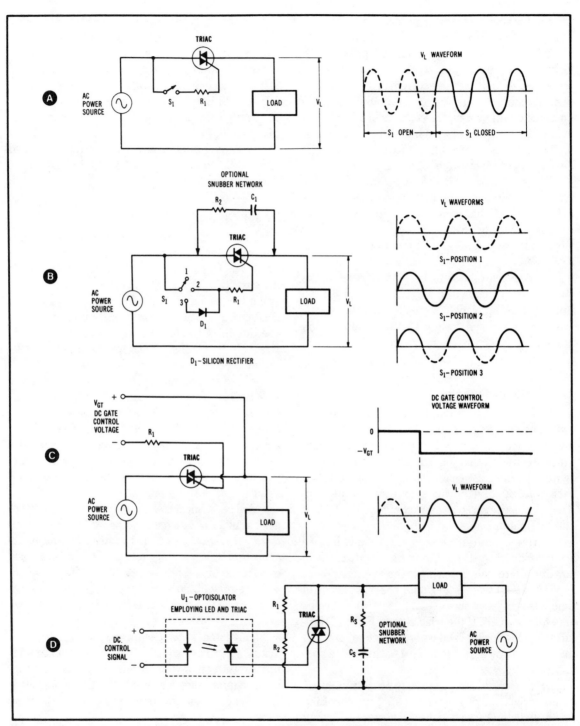

Fig. 3-12. Typical triac static switching circuits. (A) Simple triac static switch. (B) Triac static switch with three modes. (C) A dc controlled triac static switch. (D) Optoisolator-triggered triac static switch.

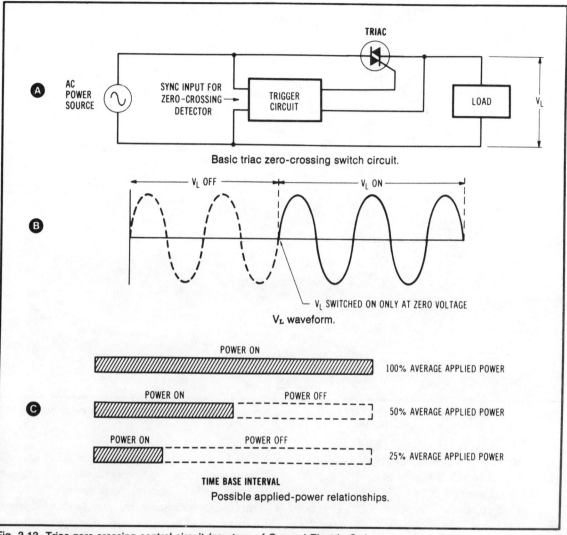

Fig. 3-13. Triac zero-crossing control circuit (courtesy of General Electric Co.).

triac gate circuits include unijunction transistors (UJTs) and special two-transistor switch configurations usually fabricated as one integrated circuit. These devices will be covered in the next chapter.

EXPERIMENT NO. 3-1, TRIAC CHARACTERISTICS

1. PURPOSE: The purpose of this experiment is to investigate the switching characteristics and operation of a typical triac.

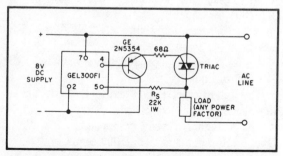

Fig. 3-14. Triac zero-crossing control circuit (courtesy of General Electric Co.).

87

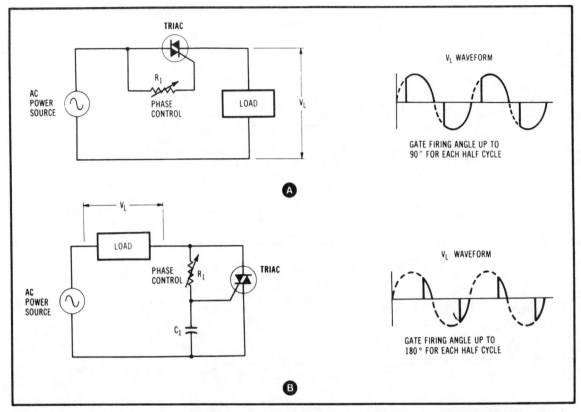

Fig. 3-15. Triac phase-delay gate circuits. (A) Simple resistive phase-delay gate circuit. (B) RC phase-delay gate circuit.

2. MATERIAL AND EQUIPMENT:

   a. Ac Power Transformer, 120/25.2 Vac, 60 Hz, 2 A: RS 273-1512 (Radio Shack) or equal.

   b. Multimeter, Analog or Digital.

   c. Battery, lantern type, 6 Vdc: Eveready 509, Ray-O-Vac 941, RS 233-006, or equal. A dc power supply can be substituted if available.

   d. Triac, 2-6 A, 200 Vdc: 2N6071, RS 276-1001, or equal.

   e. LED, Red, approximately 2 V, 20 mA: RS 271-1720, or equal.

   f. LED, Green, approximately 2 V, 20 mA: RS 271-022 or equal.

   g. Potentiometer, 5 kohms, 1 Watt, linear taper: RS 271-1714 or equal.

   h. Resistors, carbon composition, 1/2 Watt: One —270 ohms and two—1000 ohms.

   i. Switch, DPDT, toggle, 3 A/125 Vac: RS 275-602 or equal.

3. TEST PROCEDURES:

   a. Construct the test circuit given in Fig. 3-17. You can substitute similar components in this circuit provided you observe proper power levels and compute test data accordingly. Note that individual LEDs are used to indicate the direction of ac current flow. You may elect to use one of the dual color

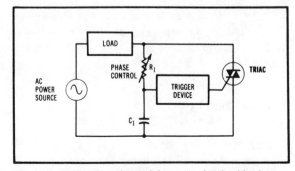

Fig. 3-16. Basic triac phase-delay gate circuit with trigger.

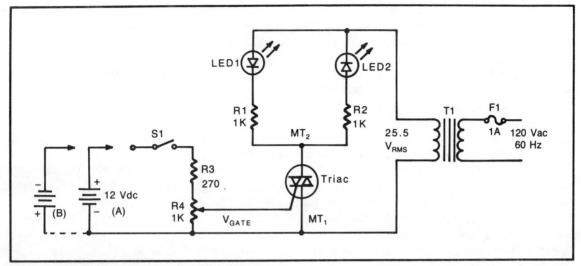

Fig. 3-17. Test circuit for basic triac characteristics.

LEDs which indicates the direction of ac current flow. The RS 276-035 Tri-Color LED (or equal) is an example of this type of LED.

b. Connect the 12 Vdc power source (A) to provide a positive gate voltage. Close S1 and slowly increase the triac gate voltage with R4 until LED1 is turned on. Measure the V_{GT}, Gate Trigger Voltage, for Triggering Mode I (MT_2 (+) and V_{GT} (+)). Record this data in Table 3-1.

c. Continue to increase the gate voltage until LED2 is turned on. This allows you to measure V_{GT}, Gate Triggering Voltage, for Mode IV ($MT_2(-)$ and V_{GT} (+)). Again record this data.

d. Reverse the gate voltage polarity by switching the gate voltage supply as shown in (B) and slowly increase the triac gate voltage with R4 until LED1 is turned on. Measure V_{GT}, Gate Trigger Voltage, for Mode II ($MT_2(+)$ and $VGT(-)$). Continue to increase the gate voltage until LED2 is turned on. Measure V_{GT}, Gate Trigger Voltage, for Mode III ($MT_2(-)$ and $V_{GT}(-)$). Record the data for both modes.

e. Switch the multimeter to a low dc milliampere range (about 0-50 mAdc) and connect into the test circuit to measure I_G. Repeat the basic procedures in Steps b. through d., above, to measure I_{GT}, Gate Trigger Current, for the four modes of operation. Record this data in Table 3-1.

4. ANALYSIS OF TEST DATA:

a. The test data shows the required values of V_{GT} and I_{GT} for the triac used in the test circuit. You should compare this data with the manufacturer's specifications, if available. If a 2N6071 or an RS 276-1001 is used in the experiment, the values of V_{GT} and I_{GT} will be in the neighborhood of 2 Vdc and 20 mA, respectively. If a sensitive gate triac, such as a 2N6071A or 2N6071B, is used in the test circuit, the gate trigger current will be on the order of 3 to 5 mAdc. These lower values are useful when digital logic devices, such as the 7400 TTL family, is employed to control the triac for high load voltages up to 120 Vac or higher.

Table 3-1. Triac Test Data.

Gate trigger Mode	V_{GT} Volts	I_{GT} mA
Mode I: MT_2 (+) V_G (+)		
Mode II: MT_2 (+) V_G (−)		
Mode III: MT_2 (−) V_G (−)		
Mode IV: MT_2 (−) V_G (+)		

b. The test circuit for the triac employed an ac power source. Thus, the triac is turned *off* at the end of each ac half cycle. However, if a dc power source is used for the load current (MT_2 to MT_1 current), the triac will operate in a manner similar to two SCRs connected in a full-wave control circuit. Like the SCR, the triac remains in a state of conduction after the gate trigger voltage is removed or reduced to zero. The triac cannot be turned off until the load current is reduced below minimum holding current, I_H. You can refer to Experiment 2-1 in Chapter 2 for test procedures to determine the triac's minimum holding current. I_H for the 2N6071 is on the order of 50-70 mA. Also, you may be interested in conducting a complete series of tests to determine the four modes of operation using a dc power supply.

EXPERIMENT NO. 3-2, TRIAC AC OPERATION

1. PURPOSE: The purpose of this experiment is to use three circuits to investigate the operation of triacs in ac control. These circuits cover simple half-wave and full-wave types of operation. Although the control circuits use low power supply voltages (for safety considerations) and LEDs to represent the loads, you can scale these control circuits for operation at 120 Vac with lamps, heaters, or motors for the load circuits. However, pay close attention to component power ratings and safety considerations when working with 120 Vac power.
2. MATERIALS AND EQUIPMENT:
    a. Ac Power Transformer, 120/25. Vac, 60 Hz, 2 A: RS 273-1512 (Radio Shack) or equal.
    b. Multimeter, Analog or Digital.
    c. Oscilloscope (Optional), Dual trace, general purpose (frequency response of 5 MHz or higher). Any modern oscilloscope marketed by such companies as B&K Precision Products Group, Heath Company, Leader Instrument Company or VIZ Manufacturing will be satisfactory for these experiments.
    d. Triac, 2-6 A, 200 Vac: 2N6071, RS 276-1001, or equal.
    e. Capacitor, $2.0\mu F/50$ Vdc: Two RS 272-1055 connected in parallel or equivalent capacitors.
    f. Capacitor, $470\mu F/50$ Vdc: RS272-1046 or equal.
    g. Diode Rectifier, 1 A/200 PIV: 1N4003, RS 272-1102, or equal.
    h. Fuse, 1 A/120 Vac, with In-Line Fuse Holder: RS 270-1273 and RS 270-1281, or equal.
    i. LED, Tri-Color, approximately 2 V, 25 mA: RS 276-035 or equal.
    j. Resistors, carbon composition, 1/2 Watt: 100 ohms (1), 220 ohms (2), 470 ohms (2). Also 1 watt: 820 ohms (2).
    k. Switch, SPST, momentary contacts, normally open: RS 275-609 (package of two) or equal.
3. TEST PROCEDURES:
    a. Construct the test circuit given in Fig. 3-18. Note that Switch S1 is not listed in the materials section—you can use a three position switch or simulate the function by merely connecting the wiring to achieve the required switching functions.
    b. Apply power to the transformer with S1 in Position 1. The LED should be *off*. Switch to Position 2. The LED glows red, indicating the flow of load current in only one direction. When S1 is switched to Position 3, load current is conducted in *both* directions through the triac and the LED should glow yellow. Use an oscilloscope to analyze the gate switching and load current waveforms.
    c. The circuit in Fig. 3-18 provides a three-way power switching capability for *off, half-wave* and *full-wave* operation. This allows power to the load to be adjusted to *no-load, half-load* and *full-load* conditions. By varying the value of R2 to larger values, the firing angle can be delayed to provide *one-fourth load* to approximately *full-load* conditions. Although this is a simple and reliable power control circuit, it does possess one serious problem, radio frequency interference (rfi). The rfi is generated since the triac is not switched on until the ac voltage source reaches several volts in amplitude—this produces a sudden turn-on of current, and thus the rfi. You can detect this rfi by placing a small AM transistor radio near the control circuit and tuning the radio to a clear frequency (i.e., tune *away* from any AM station).

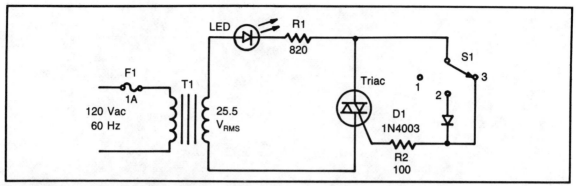

Fig. 3-18. Three-way simple ac triac control circuit.

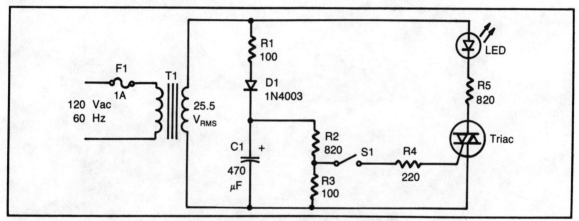

Fig. 3-19. Triac power control circuit using self-derived dc gate switching voltage.

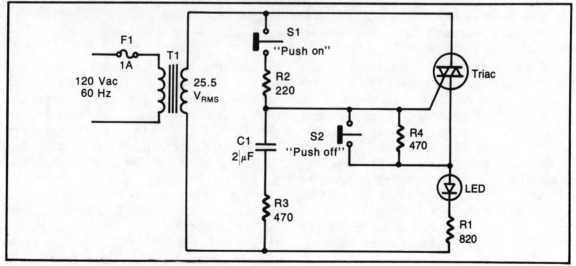

Fig. 3-20. Self-latching triac ac control circuit.

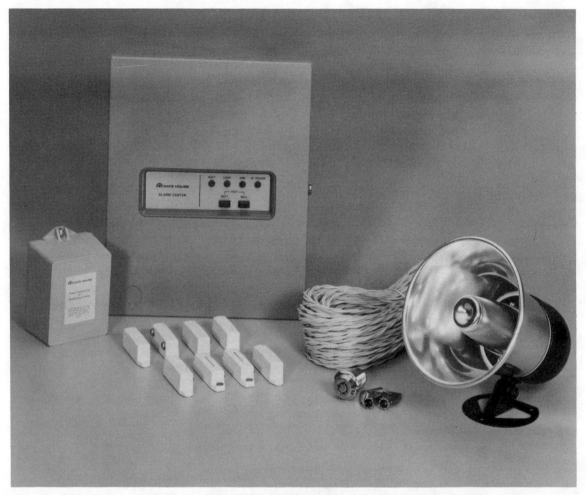
Fig. 3-21. Typical home security alarm system (courtesy of Radio Shack, Division of Tandy Corp.).

d. Rfi generated in triac power control circuits can be virtually eliminated by using gate trigger circuits that turn on the triac at the *beginning* of the ac power cycle. For example, a constant dc gate trigger voltage allows the triac to be switched on at the beginning of the ac cycle—this eliminates any abrupt switching which generates rfi radiation. Figure 3-19 shows how this dc trigger voltage may be developed from the ac power source. Build this circuit and test for presence of rfi. Observe that this circuit provides for either *no-load* or *full-load* conditions—no intermediate power loads are possible.

e. Sometimes a momentary contact is required to turn on or off a triac control circuit. Some security alarm systems use this principle to detect unauthorized intrusion. Figure 3-20 illustrates this concept. Connect the self-latching control circuit provided in Fig. 3-20. Observe that momentary switching at S1 causes the triac to be switched *on* (the LED glows yellow). The initial gate-switching pulse is provided through R2 as S1 is momentarily closed. As the triac is turned on, C1 is charged during one of the ac half-cycles. C1 starts to discharge through R3 when the triac is turned on. However, the slow time constant of R3/C1 maintains a voltage to the triac gate terminal after the ac power half-cycle—this turns on the triac during the next ac half-cycle. At the same time, the capacitor is recharged for the next ac half-cycle. Thus, the triac continues to be turned on as it is *self-latching*. In order to turn on the triac, Switch S2 is closed momentarily. Closing S2 has the effect of shorting the triac gate terminal to load terminal MT_1. Thus, a zero gate voltage condition causes the triac to be turned off.

A typical home security alarm system for easy installation is shown in Fig. 3-21. These systems use either SCRs with dc power supplies, or triacs with ac power supplies. The dc powered systems usually have a battery for sustained operation in the event the ac power source is cut off.

Chapter 4

Other Thyristor Devices

IN CHAPTERS 2 AND 3, WE COVERED THE SCR and triac, the primary workhorses in the thyristor family. Almost all ac and dc power-switching requirements can be handled with these two versatile devices.

Trigger circuits for SCRs and triacs employ a wide variety of trigger or gate-switching devices. Since the SCR was introduced in the late 1950s, gate-switching techniques and devices have been in a constant state of development. Multilayer thyristor research and advances in technology have resulted in a sophisticated family of two-, three-, and four-terminal devices. Variations in fabricating pn junctions have made possible the production of devices with varied characteristics and operational capabilities. We will examine some of these interesting devices in this chapter.

The present trend in the ac and dc power control and switching field appears to be focusing on the use of optical or optoisolator triggering employing digital and/or automated control. The introduction of the microprocessor, and related digital control circuits, has opened a new and promising era. Many of the devices covered in this chapter will be used in automated control systems. Others will be used as replacements items in existing control systems.

SHOCKLEY DIODE

The *Shockley diode,* a two-terminal semiconductor, is the most elementary of the pnpn structures. It is designed for unilateral, or half-wave operation. Since it usually is produced as a low-current, low-voltage switching device, the Shockley diode is used in SCR trigger circuits.

Theory of Operation

Figure 4-1 shows basic details of the Shockley diode. As with the SCR, the pnpn structure of the Shockley diode may be analyzed in terms of two equivalent transistors, $p_1 n_1 p_2$ and $n_1 p_2 n_2$. Unlike the SCR, this device does not employ a gate connection. The $n_1 p_2$ junction is reverse biased in conventional Shockley diode circuits.

The Shockley diode operates in the following

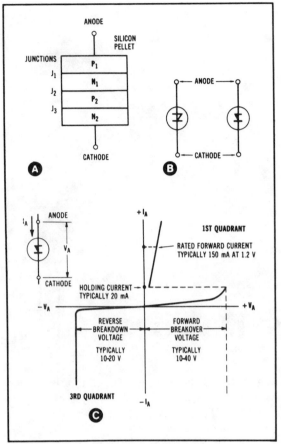

Fig. 4-1. The Shockley diode. (A) Basic construction. (B) Common symbols. (C) VI characteristic curve.

manner. When the forward voltage is increased to the breakdown level, and avalanche current flows across the n_1 p_2 junction. This causes both equivalent transistors to "snap", or latch, into full conduction. The turn-on time, a function of both the type of device being used and the external circuit characteristics, may be as low as 100 ns. Turn-off time is typically on the order of 5 μs.

Applications

A typical SCR trigger circuit employing a Shockley diode is shown in Fig. 4-2. Note that V_{GEN}, the trigger-circuit source voltage, must be in phase with the ac power source voltage. The Shockley diode is D2. Diode D1, a conventional silicon rectifier, applies only the positive half-cycles of V_{GEN} to the trigger circuit. The R1 C1 time constant determines the charging voltage at Point A. When the voltage reaches the forward breakover level of the Shockley diode, D2 "snaps" into conduction and fires the SCR. The resulting low impedance presented by D2, R2, and the SCR gate circuit discharges the capacitor in time for the next positive half-cycle. The firing angle of the SCR can be adjusted by varying the resistance of R1.

Shockley diodes are seldom employed in new thyristor control circuit design. They are used primarily as replacement items in existing equipment.

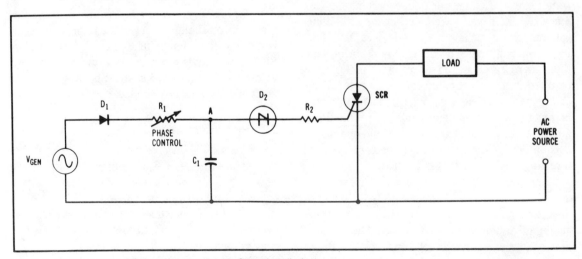

Fig. 4-2. Phase control SCR trigger circuit with Shockley diode.

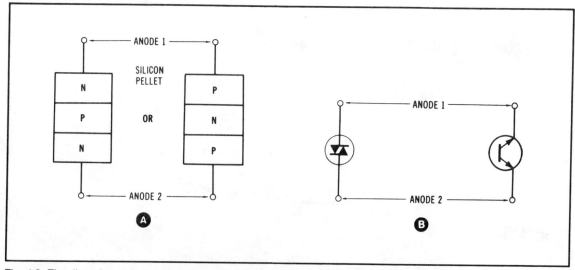

Fig. 4-3. The diac trigger diode. (A) Basic npn or pnp construction. (B) Common symbols for the diac.

DIAC

Sometimes referred to as a trigger or bidirectional diode, the diac (Fig. 4-3) is a two-terminal semiconductor capable of blocking or conducting current in either direction. In its most elementary form, the diac is a three-layer symmetrical npn silicon structure. The diac is used extensively as a trigger device for triac gate circuits.

Theory of Operation

The diac resembles an npn bipolar transistor with no base connection. Unlike the bipolar transistor, the diac possesses uniform construction; N-type and P-type doping is essentially the same at both junctions. The diac may be constructed as either an npn or pnp structure.

Electrically, the diac structure provides for symmetrical circuit operation. Figure 4-4 shows a typical VI characteristic curve. The presence of either positive or negative voltage across the two diac terminals results in one forward-biased and one reverse-biased pn junction. At low voltage levels, the reverse-biased junction prevents any appreciable current; only a small leakage current flows through the device. When the applied voltage is increased to the breakover voltage level (V_{bO}), the reverse-biased junction proceeds into avalanche breakdown conduction. At this point, the device exhibits negative resistance characteristics; current conduction increases while voltage across the device decreases.

Diac turn-on time varies with the type of device and the external circuit being used. Typical turn-on time ranges from about 50 to 500 ns. Turn-off time for diac circuits is much longer, ranging up to about 1000 ns.

Most diacs have a power dissipation of about 300 mW to 1 W. Peak trigger currents for short periods of time range up to about 2 A. Switching-voltage symmetry is usually rated at ± 2 to ± 4 V.

Applications

Diacs are reliable, economical trigger devices for thyristor control applications. They are used extensively as triggers for triac ac control circuits for lamp dimmers, heating controls, motor-speed controls, and similar applications. Light-activated diacs are incorporated in some optoisolator devices.

Figure 4-5A shows a typical triac phase-control circuit using a diac as a trigger diode for the triac. One application for this type of circuit is the common lamp dimmer shown in Fig. 4-5B. This device is used for controlling the intensity of incandescent lamps.

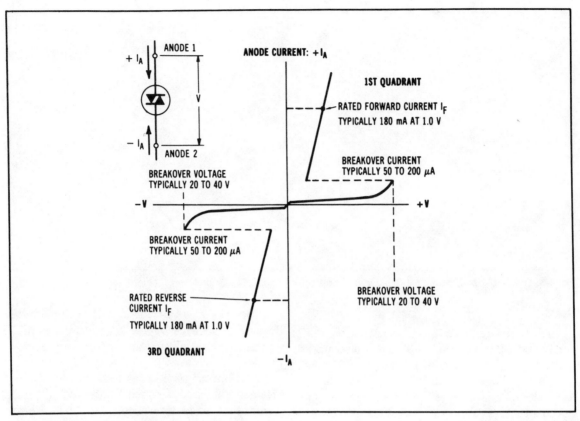

Fig. 4-4. Typical diac VI characteristic curves.

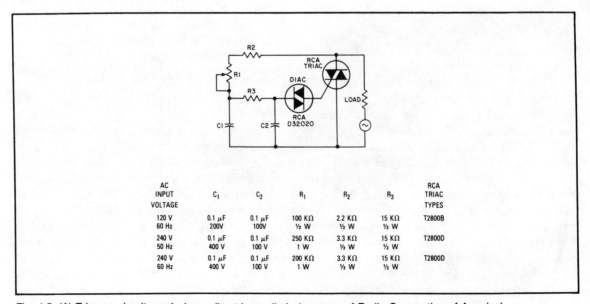

Fig. 4-5. (A) Triac ac circuit employing a diac trigger diode (courtesy of Radio Corporation of America).

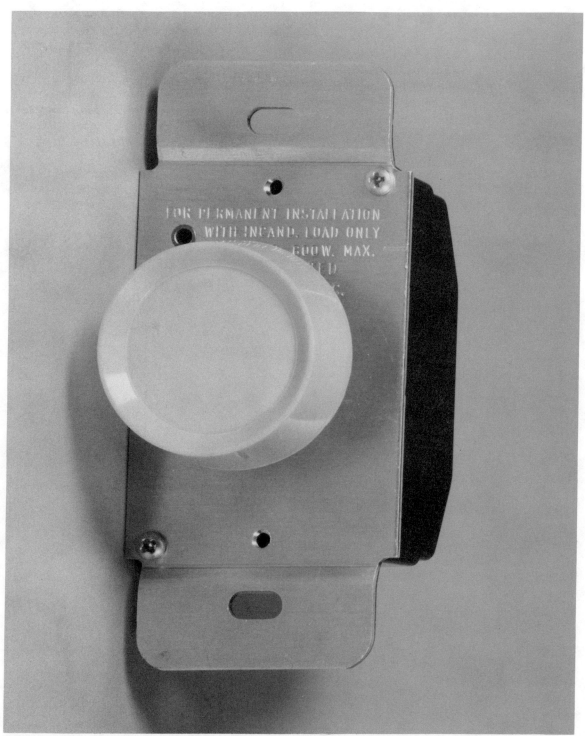
(B) Typical lamp dimmer employing a thyristor device to control (courtesy of Radio Shack a Division of Tandy Corporation).

SILICON-CONTROLLED SWITCH

Often referred to by the abbreviation SCS, the *silicon-controlled switch* is a four-terminal, four-layer pnpn device used in low-power switching applications. Similar to the SCR in construction, the SCS has two gate terminals. Either gate can be used to turn on or turn off the main current through the device. Details of the SCS are given in Fig. 4-6.

Theory of Operation

An equivalent two-transistor switch configuration can be used to describe the operation of the SCS. The only difference between this and the SCR two-transistor model is in the use of a second gate terminal. Figure 4-7 shows the SCS two-transistor model connected to a load and to a dc power supply.

Prior to the application of a trigger signal to either gate terminal, the SCS is turned off. The collector-base junction of both transistors are reverse-biased, preventing current through the device. Note that a dc continuity path from the anode gate to ground is not permissible, since this would allow load current to flow through the forward-biased emitter-base junction of Q1.

A positive cathode-gate trigger signal (V_{GTC}) or a negative anode-gate trigger signal (V_{GTA}) will turn on the SCS. Either gate signal results in base current in the respective transistor, thereby resulting in collector current. Since this current

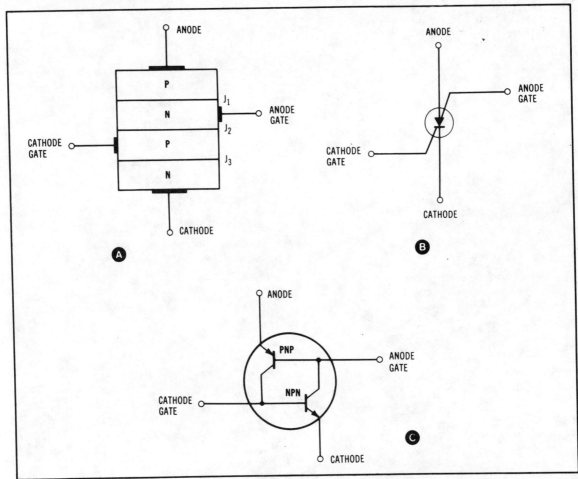

Fig. 4-6. The silicon controlled switch (SCS). (A) Basic construction. (B) Symbol. (C) Equivalent two-transistor model.

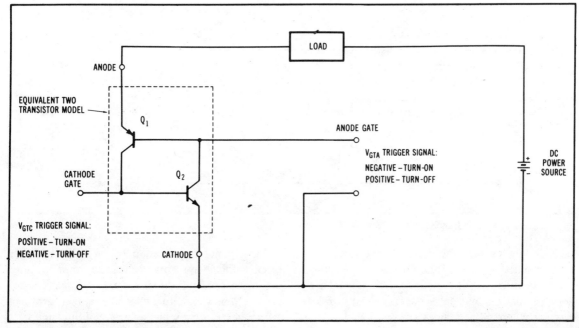

Fig. 4-7. Trigger modes for the SCS.

flows through the base of the other transistor, the regenerative action immediately latches both transistors into saturation.

Cathode-gate triggering is much more sensitive than anode-gate triggering. For example, specifications for the 3N86 silicon controlled switch list cathode-gate trigger current as low as 1 μa, while the anode-gate trigger current is 100 μa. Both anode- and cathode-gate trigger voltages are on the order of 0.8 V.

Once the SCS is in a state of conduction, there are three basic methods to turn it off—a negative trigger signal to the cathode gate, a positive trigger signal to the anode gate, or a momentarily short of the anode and cathode terminals.

The negative trigger signal applied to the cathode gate turns off Q2, the npn transistor. This, in turn, cuts off the base current to Q1, turning off the collector current of Q1. Similarly, a positive trigger signal applied to the anode gate turns off the pnp transistor. This interrupts the base current to Q2, turning it off. Note that the operation of the SCS is virtually identical to the operation of the pnp-npn transistor latching configuration. In fact, a model of the SCS can be built using matching pnp and npn transistors.

Like the SCR, the SCS can be turned off by reducing the anode-cathode current below the holding current level. As stated earlier, this can be accomplished by shorting the anode and cathode terminals with a mechanical or electronic switch. Figure 4-8 illustrates how a transistor switch can be used for this purpose. An alternate method for SCS turn-off is simply to open the anode circuit. Again, a transistor switch can be used for this function.

The turn-on time of the SCS is approximately equal to that of the SCR, typically 1.5 to 2.0 μs. Unlike the SCR, the SCS exhibits a fast turn-off time on the order of 5 μs.

Typical VI characteristics for an SCS are given in Fig. 4-9. The values of operating parameters shown on this graph are typical of the 3N81 through 3N85 series of SCS devices. Figure 4-10 presents the specification sheets for the 3N83 SCS.

Applications

The SCS may be used in many applications

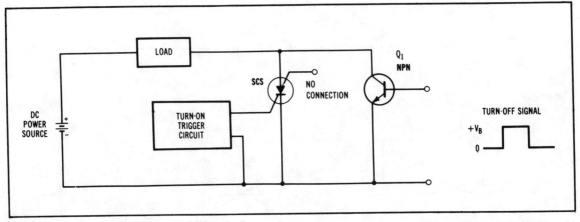

Fig. 4-8. Use of an npn transistor for SCS turn-off.

where a low-level SCR with gate turn-on and turn-off capability is required. Lamp and relay drivers, digital counters, shift registers, voltage-level sensors, oscillators, and pulse generators are typical circuits where SCS devices are employed.

One popular application for the SCS is in driver circuits for neon display tubes. Large neon tubes are often used where an electronic display is required to be visible to a widely dispersed group of personnel. For example, a display of time in hours,

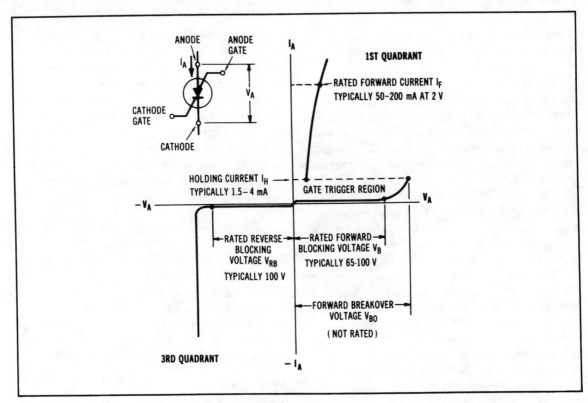

Fig. 4-9. Typical silicon controlled switch VI characteristics.

Silicon Controlled Switch

3N83

The General Electric Type 3N83 is PLANAR PNPN switch with separate leads provided to each of its four semiconductor regions to form the equivalent of an integrated PNP-NPN complementary transistor pair. It has been characterized as a low-cost, latching-type driver for Nixie tubes, alphanumeric display tubes and neon lamps. As such, it is ideally suited for very simple counter circuitry and applicatins requiring gate turn-off as well as gate turn-on. Special features of the 3N83 neon driver include its ability to operate independently of the changes in ionization and deionization time of the lamp and its total freedom from inadvertant triggering caused by line transients (dv/dt).

The 3N83 is housed in a four-leaded TO-18 size package. All junctions are completely oxide passivated to provide maximum long term reliability. Other PNPN devices in this series provide characterization suitable for a wide variety of switching functions and are described in General Electric Publications 65.16, 65.18 and 65.19.

Features:

- Latching driver for neon lamps
- Design parameters specified at worst-case temperatures
- Gate turn-on and turn-off
- Eliminates lamp ionization and deionization-time problems
- All planar, completely oxide passivated
- TO-18 size

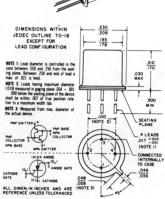

PNPN NEON DRIVER CHARACTERIZATION
THEORY OF OPERATION

Considering a PNPN device as an integrated circuit in which an NPN transistor drives the lamp load and a PNP transistor provides the NPN with latching characteristics results in a family of outstanding driver circuits.

Figure 1 compares the SCR and integrated circuit symbols of a PNPN.

Figure 2 illustrates operation in a typical circuit. Initially, assume the NPN transistor is reverse biased. Its collector voltage (pin 3) will be at the collector supply voltage minus the lamp extinction voltage. Typically this will be 50 volts, which reverse biases the PNP transistor. The lamp is extinquished.

Applying a positive trigger pulse saturates the NPN transistor lowering its collector voltage to within 1.0 volt of ground. This forward biases the PNP transistor turning it on and supplying additional base drive to the NPN transistor. The 3N83 is characterized so that the NPN transistor remains saturated even after the triggering input is removed. The minimum PNP emitter current to guarantee this is defined as the holding current (I_H).

To turn off the 3N83 it is only necessary to interrupt the PNP emitter current momentarily. This removes the NPN base drive allowing the collector to rise to the lamp extinction voltage reverse biasing the PNP. The 3N83 is also characterized to turn off by a negative pulse at the NPN base. This is equivalent to gate turn-off of an SCR. The negative pulse diverts all the PNP collector current to reverse bias the NPN thus turning off the driver.

The advantages of this circuit include (a) latching characteristics (b) no rate effect problems because the PNP emitter is normally reverse biased (c) spikes and ripple on either power supply will not turn on the driver accidentally (d) turn-off is readily achieved in the low voltage, low power PNP emitter or NPN base circuits (e) lamp ionization and deionization characteristics do not affect the circuit (f) dissipation is low since I_H is generated from a low voltage supply and (g) the circuit is low cost.

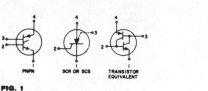

FIG. 1

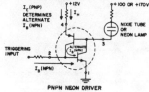

FIG. 2

Fig. 4-10. Specifications for a typical silicon controlled switch (courtesy of General Electric Co.). (Continued through page 106.)

3N83

absolute maximum ratings: (25°C) (unless otherwise noted)

		PNP[1]	NPN[1]	
Voltages				
Collector to Emitter (R = 10K)	V_{CER}	−70	70	volts
Collector to Base	V_{CBO}	−70	70	volts
Emitter to Base	V_{EBO}	−70	5	volts
Emitter Current				
Continuous DC[2]	I_E	50	−50	ma
Peak Recurrent ($T_A = 100°C$, 100 μsec. pulse width, 1% duty cycle)	$I_{E(PEAK)}$	0.1	−0.1	amp
Peak Non-Recurrent (10 μsec. pulse width)	$I_{E(PEAK)}$	0.5	−0.5	amp
Collector Current[3]				
Continuous DC[2]	I_C	−10	25	ma
Peak Recurrent ($T_A = 100°C$, 100 μsec. pulse width, 1% duty cycle)	$I_{C(PEAK)}$	−50	50	ma
Dissipation				
Total Power[2]	P_T	200	200	mw
Temperature				
Operating Junction	T_J	−55 to +125		°C
Storage	T_{STG}	−65 to +200		°C

---

electrical characteristics: (25°C) (unless otherwise specified)

		PNP[2]		NPN[2]		
		Min.	Max.	Min.	Max.	
DC CHARACTERISTICS						
Collector to Emitter Breakdown Voltage						
($I_C = 0.1$ μa)	BV_{CEO}	−70				volts
($I_C = 0.1$ μa, R = 10 KΩ)	BV_{CER}			70		volts
Collector to Base Breakdown Voltage ($I_C = [\pm]^4$ 0.1 μa, $I_E = 0$)	BV_{CBO}	−70		70		volts
Emitter to Base Breakdown Voltage ($I_C = 0$, I_E [NPN] = 20 μa, I_E [PNP] = −1 μa)	BV_{EBO}	−70		5		volts
Collector Saturation Voltage ($I_C = 25$ ma, $I_B = 2.5$ ma)	$V_{CE(SAT)}$[5]				1.2	volts
Base Saturation Voltage ($I_B = 1$ ma, $I_C = 5$ ma)	$V_{BE(SAT)}$				0.9	volts
Forward Current Transfer Ratio ($V_{CE} = 0.5$ V, $I_C = 3$ ma)	h_{FE}			15		
($V_{CE} = -2.0$ V, $I_C = 1.0$ ma)	h_{FE}	0.1				
CUTOFF CHARACTERISTICS						
Collector to Emitter Leakage Current ($V_{CE} = -70$ Vdc, $T_A = 125°C$)	I_{CEO}		−20			μa
($V_{CE} = 70$ Vdc, $R_B = 10K$ Ω, $T_A = 125°C$)	I_{CER}				20	μa
Collector to Base Leakage Current ($V_{CB} = [\pm]^4$ 70 Vdc, $I_E = 0$, $T_A = 125°C$)	I_{CBO}		−20		20	μa
Emitter to Base Leakage Current ($V_{EB} = -70$ Vdc, $I_C = 0$, $T_A = 125°C$)	I_{EBO}		−20			μa
($V_{EB} = 5$ Vdc, $I_C = 0$)	I_{EBO}				20	μa
COMBINED DEVICE CHARACTERISTICS		Min.	Max.			
Collector Capacitance[6] ($I_E = 0$, $V_{CB} = 20$ Vdc)	C_{ob}		20			pf
Forward Voltage ($I_E = 50$ ma)	V_F		1.4			volts
($I_E = 50$ ma, $T_A = -55°C$)	V_F		1.9			volts
Holding Current (See test circuit below)	I_H		4.0			ma
Current to Trigger (See test circuit below)	$I_{TRIGGER}$		150			μa
Voltage to Trigger (See test circuit below, $T_A = -55°C$)	$V_{TRIGGER}$		1.1			volts
(See test circuit below, $T_A = 125°C$)	$V_{TRIGGER}$	0.2				volts
Turn-On Time (See test circuit below, I_B [NPN] = 0.4 ma)	t_{on}		1.5			μsec.
Turn-Off Time (See test circuit below, $V_F = 0$ during recovery)	t_{off}		8			μsec.
(See test circuit below, $T_A = 125°C$)	t_{off}		15			μsec.

notes:

1. All individual transistor characteristics are given with the emitter of the complementary transistor open; e.g., for $V_{CE(SAT)}$ (NPN): I_C (NPN) = 25 ma, I_B (NPN) = 2.5 ma, I_E (PNP) = 0.
2. Derate current and power linearly to zero at 125°C. The absolute maximum rating at any given temperature shall be in terms of the more conservative of the two parameters, i.e., current or power.
3. The collector current of the PNP is identical to the base current of the NPN and the collector current of the NPN is identical to the base current of the PNP. (See outline drawing.)
4. ± indicates that appropriate polarity should be chosen for transistor under test.
5. $V_{CE(SAT)(NPN)}$ is modulated by the PNP emitter current.
6. Collector capacitance is measured between terminals 2 and 3. (See outline drawing.)

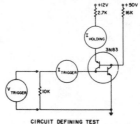

CIRCUIT DEFINING TEST CONDITIONS

FIG. 3

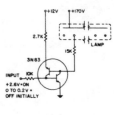

TYPICAL LATCHING LAMP DRIVER

FIG. 4

522

dc characteristics

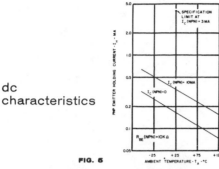

FIG. 5

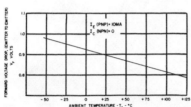

FIG. 6

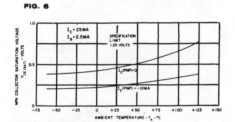

FIG. 7

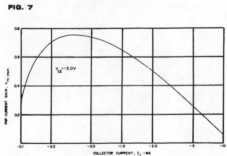

FIG. 8

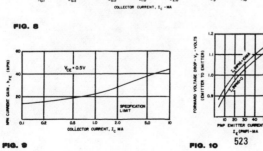

FIG. 9 FIG. 10

3N83

dynamic characteristics

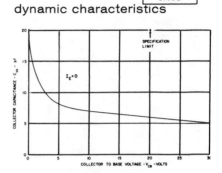

FIG. 11

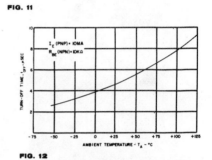

FIG. 12

triggering characteristics

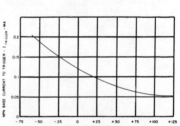

FIG. 13

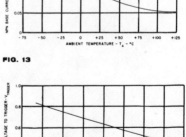

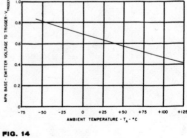

FIG. 14

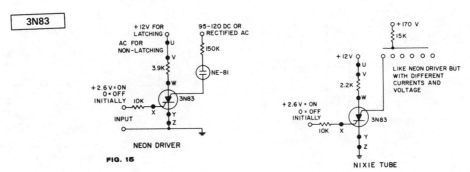

FIG. 15 — NEON DRIVER

FIG. 16 — NIXIE TUBE

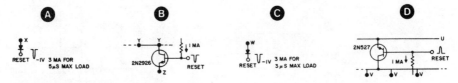

reset methods

Four common reset methods are illustrated. The driver circuit is modified by connecting or inserting the turn-off circuitry to similarly lettered points. Method (A) reverse biases the NPN base. Method (B) illustrates that several stages can be turned off simultaneously by open circuiting the NPN emitter. Method (C) reverse biases the PNP emitter while method (D) open circuits one or more PNP emitters. In each case, rate effect problems are non-existent and no special care is required in shaping the reset waveforms.

FIG. 17

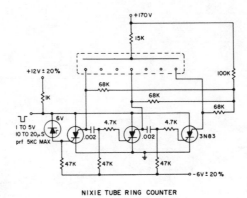

NIXIE TUBE RING COUNTER

FIG. 18

minutes, and seconds is required in many real-time analytical or processing operations. Figure 4-11 shows how the SCS can be used for this application. An interesting aspect of this circuit is that the anode *gate* terminal conducts the neon load current. The circuit in Fig. 4-11 turns on or off the 0 digit contained in the neon tube. Nine additional circuits are required to control turn-on of digits 1 through 9 at other required times.

GATE TURN-OFF THYRISTOR

The *gate turn-off* (GTO) *thyristor*, or switch, is a three-terminal device similar in construction and operation to the SCR. However, as the name implies, the GTO can be turned *off* by a reverse gate turn-on current. Thus the GTO combines the turn-on/turn-off features of a bipolar power transistor with a high-current/high-voltage characteristics of the SCR. Unlike the power transistor which requires a constant base current for turn-on, the GTO

quires only a momentary pulse for turn-on and turn-off operation.

Figure 4-12A provides some of the basic details concerning the GTO. The two-transistor regenerative circuit analogy given for SCR operation in Chapter 2 is also applicable for GTO operation. Details of GTO construction are shown in Fig. 4-13.

Operation of the GTO features high-speed gate turn-on and turn-off times, typically on the order of a microsecond or less. New GTO devices such as the Motorola MGT01000/1200/1400 series, are rated as high as 18 A of load current with blocking voltage ratings from 1000 to 1400 volts. These GTOs have faster switching times than conventional SCRs, with a typical turn-on time of only 0.6 μsec. Furthermore, these GTOs require only a 300 mA trigger current, have high surge current capabilities up to 200 A, and display low forward conduction losses at relatively high anode currents. The Motorola MGTO1000/1200/1400 series

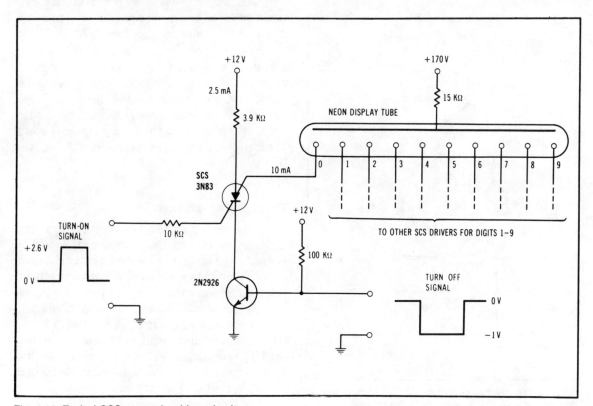

Fig. 4-11. Typical SCS neon-tube driver circuit.

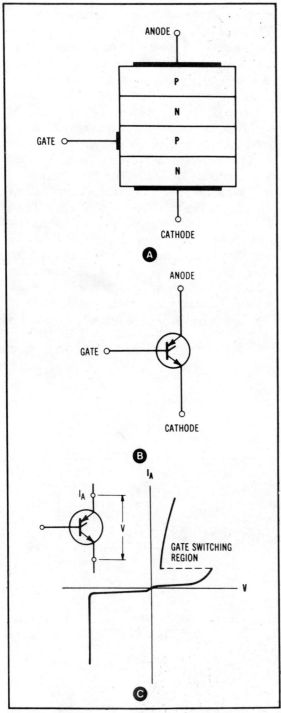

Fig. 4-12. The gate turn-off switch (GTO). (A) Basic construction. (B) Symbol. (C) Generalized VI characteristics.

devices are available in standard TO-220 packages and TO-220 overmold packages with quick disconnect fast-on terminals.

The GTO is employed in automotive electronics, switching power supplies, motor drive, and other power control applications. In addition to the Motorola MGT01000/1200/1400 series, the RCA G4000 and G4001 series of 5- and 10-amp GTO silicon controlled rectifiers are typical of the available commercial devices.

UNIJUNCTION TRANSISTOR

A unique member of the thyristor family, the *unijunction* or PN *transistor* (UJT) is a two-layer pn device with three terminals. Similar to the bipolar transistor, the UJT is fabricated on an N-type silicon bar with ohmic contacts for the two base terminals (Fig. 4-14A). The emitter section, a P-type material, is deposited between the base 1 and base 2 regions. Figure 4-14 shows details of basic construction of the device, an equivalent electrical circuit, and the commonly used symbol.

Possessing only one pn junction, the UJT is a versatile semiconductor device that exhibits negative resistance characteristics. This means that an increasing emitter current results in a decreasing voltage between the emitter and base 1 terminals. This electronic switching action is useful in generating gate-trigger signals for higher-power thyristors.

Theory of Operation

The operation of the UJT can best be explained in terms of the equivalent electrical model illustrated in Fig. 4-15. This model is connected in a test circuit with a variable dc biasing supply for the emitter terminal and fixed dc power supply for the base terminals. Resistors R1 and R2 are simply current-limiting resistors for this particular circuit. Note that this test circuit is used only to show the UJT operating characteristics; other than that, it cannot perform any useful work. We will examine a practical UJT circuit later.

The N-type silicon semiconductor region com-

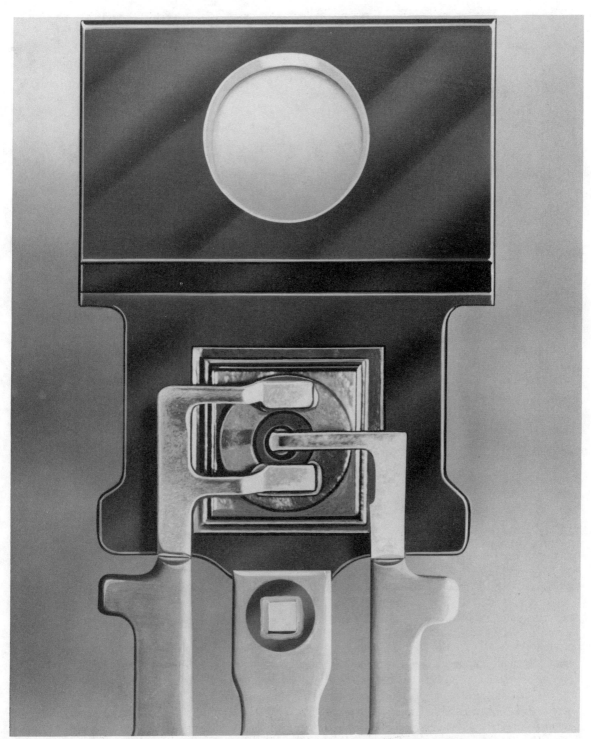

Fig. 4-13. GTO details (courtesy of Motorola Semiconductor Products, Inc.). (A) Basic GTO construction.

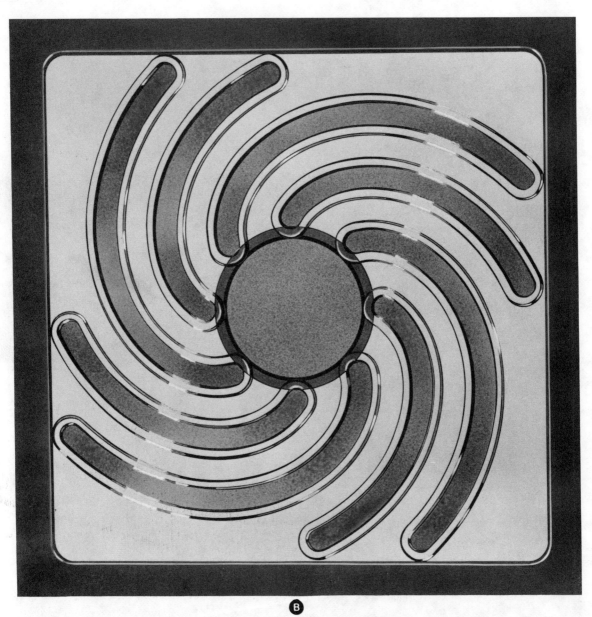

Fig. 4-13. (B) GTO emitter die geometry to aid current turn-on spreading and reduce dI/dt turn-on failure.

prising the base material is lightly doped. The resulting small number of free electrons will support only a small current between the two base terminals. With no emitter current, the dc resistance across this N-type material from base 1 to base 2 is approximately 4700 ohms to 9000 ohms. As shown in the equivalent circuit, the pn junction acts as a silicon diode connected to the two base regions. Note that the relative values of R_{B1} and R_{B2} depend on where the P-type emitter material is located along the N-type bar.

An important UJT operating parameter is the intrinsic standoff ratio. Most often referred to as η (Greek letter eta), this is defined as the ratio of

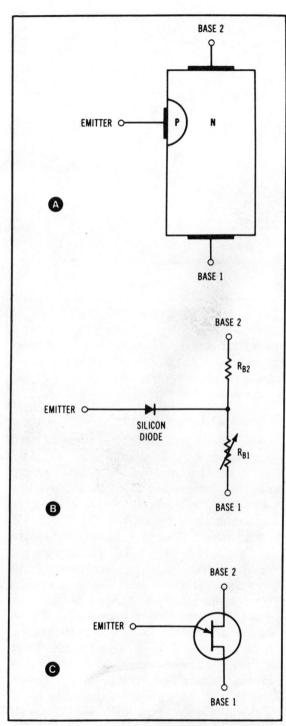

Fig. 4-14. The unijunction transistor (UJT). (A) Basic construction. (B) Equivalent electrical circuit. (C) Symbol.

R_{B1} to the total base resistance when the emitter current is equal to zero. This relationship is expressed as:

$$\eta = \frac{R_{B1}}{R_{B1} + R_{B2}} \quad \text{(Eq. 4-1)}$$

where,
η is a ratio, and
R_{B1} and R_{B2} are the equivalent base resistance in ohms.
Typical values of η vary from about 0.5 to 0.8 for most UJTs.

Here is why the intrinsic standoff ratio is so important. Knowing η and the applied voltage across the base terminals (V_{BB}), we can calculate the internal voltage across R_{B1}. This is given by:

$$V_{RB1} = \frac{V_{BB} \times R_{B1}}{R_{B1} + R_{B2}} \quad \text{(Eq. 4-2)}$$

or

$$V_{RB1} = \eta\, V_{BB} \quad \text{(Eq. 4-3)}$$

For example, if $\eta = 0.65$ and $V_{BB} = 10$ V, then

$$V_{RB1} = 0.65 \times 10 = 6.5 \text{ V}$$

Voltage V_{RB1} represents a reverse bias voltage on diode D1. In order for an emitter current (I_E) to flow, the emitter voltage (V_E) must rise above V_{RB1} by about 0.7 V, the internal barrier potential for a silicon diode. The emitter voltage that will cause the diode to be forward biased and conduct an emitter current is usually designated V_P. Accordingly, V_P can be calculated by the following equation:

$$V_P = \eta\, V_{BB} + V_D \quad \text{(Eq. 4-4)}$$

or

$$V_P = \eta\, V_{BB} + 0.7 \quad \text{(Eq. 4-5)}$$

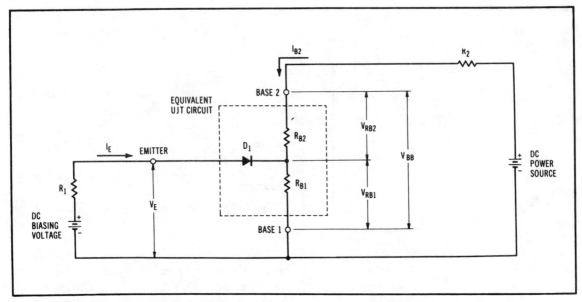

Fig. 4-15. Basic UJT circuit operation.

In the previous example, it can be shown that V_P should be equal to approximately 7.2 V.

When the emitter voltage is increased above V_P, V_{RB1} might be expected to increase accordingly. With individual components (a silicon diode and two resistors), this is exactly what would happen. However, the UJT gate voltage *decreases* as the emitter current *increases*. Figure 4-16 shows a typical UJT emitter current-voltage curve for V_{BB} adjusted to 10 V. Reducing V_{BB} will result in corresponding lower values of V_P, $V_{E(sat)}$, V_V, and I_V.

Here is the theory behind the sudden decrease in V_E as the emitter current increases. The forward biased pn junction injects a flood of charge carriers (holes, in this case) into the lightly doped N-type base region. Being positive charge carriers, they are swept toward the negative potential at base 1. This action reduced the effective resistance of R_{B1} to a value about 50 ohms. Thus, the low resistance produces a small voltage drop. Accordingly, V_E will drop to a value of less than one volt in most circuits. The UJT is now triggered to the maximum current through the base regions. This is sometimes referred to as the *on-state condition*.

The base 2 region is virtually unaffected by the emitter current; resistance R_{B2} remains approximately the same. Thus the primary action of the unijunction transistor takes place in the emitter and R_{B1} portions of the device.

Typical UJT Specifications

Unijunction transistors are available with power dissipation ratings up to about 450 mW. Typical interbase and emitter reverse voltage ratings range from about 30 V to 60 V. Maximum rms peak emitter current levels are on the order of 50 mA and 2 A, respectively.

The emitter triggering current (I_P) ranges from about 0.4 μA to 12 μA. This characteristic can be used in high-impedance sensing circuits to detect extremely small levels of current.

Triggering (or turn-on) time of UJTs is seldom specified by manufacturers. In general, the UJT is useful in oscillator and timing circuits for frequencies ranging from 1 Hz to 1 MHz. Emitter-voltage fall time is sometimes specified for relaxation-oscillator circuits as a function of capacitance. These values usually range from a few microseconds to about 100 μs. Additional information concerning UJT specifications is contained in Fig. 4-17.

Applications

The UJT has been employed in a wide variety of circuits involving electronic switching and voltage or current sensing applications. These include triggers for thyristors, oscillators, pulse and sawtooth generators, timing circuits, regulated power supplies, and bistable circuits.

Figure 4-18 shows one of the most common UJT circuits, the basic *relaxation oscillator*. Voltage waveforms for two key points in the circuit illustrate the relationships between emitter and interbase circuit operation.

When S1 is initially switched on, a capacitor charging current flows through the series R1 C1 network. At time zero, or t = 0, the charging capacitor exhibits a short circuit, and the emitter voltage is approximately zero volts. Governed by the R1 C1 time constant, the voltage across the capacitor starts to rise as the capacitor charges. The increasing voltage is illustrated in the emitter-voltage waveforms in Fig. 4-18B.

As in the increasing capacitor voltage reaches V_P, the UJT is suddenly switched on. At this time, the internal base 1 resistance (R_{B1}) drops to a very low value, about 50 ohms. The resulting low-resistance path through the UJT (emitter to base 1) and R_3 abruptly shorts out the capacitor. Note that the *discharging time constant*, C1 ($R_{B1} + R_3$), is much smaller than the R1 C1 *charging time constant*.

When V_E drops to nearly zero volts, the emitter circuit is turned off, and the UJT is restored to its initial state. As long as S1 is closed, the UJT emitter turn-on and turn-off cycles are repeated. During each cycle, a voltage pulse is generated at the base 1 terminal. With proper UJT circuit design, the V_{BASE1} should be approximately equal to the gate trigger voltage (V_{GT}) of the thyristor. The value of V_{BASE1} in a UJT relaxation-oscillator circuit can be calculated by the following equation:

$$V_{BASE1} = \frac{R3\ V_{BB}}{R2 + R3 + R_{BB(min)}} \quad \text{(Eq. 4-6)}$$

where,
V_{BASE1} is the peak output voltage in volts developed across R3 while the UJT is in a state of full conduction,
R2 and R3 are external circuit resistors (resistances in ohms),

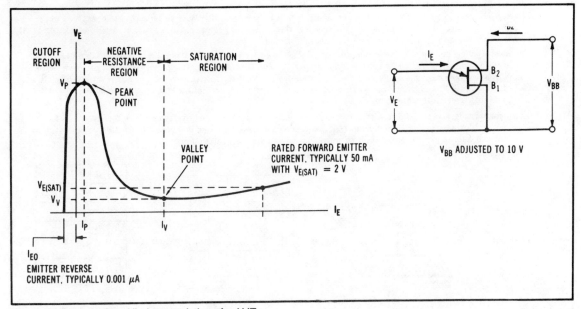

Fig. 4-16. Static emitter VI characteristics of a UJT.

Silicon Complementary Unijunction Transistor

COMPLEMENTARY UNIJUNCTION

The General Electric D5K2 Complementary Unijunction Transistor is a silicon planar, monolithic integrated circuit. It has unijunction characteristics with superior stability, a much tighter intrinsic-standoff ratio distribution and lower saturation voltage.

FEATURES

- Guaranteed stability of better than 1.0% from $-15°C$ to $+65°C$ and better than 2.0% from $-55°C$ to $+100°C$
- Low leakage current: less than 100 nA
- Ability to temperature compensate and calibrate at room temperature
- Up to 100 kHz operation

WHAT IS A COMPLEMENTARY UNIJUNCTION TRANSISTOR?

The General Electric D5K is a silicon planar passivated semiconductor device with characteristics like those of a standard unijunction transistor except that the currents and voltages applied to it are of opposite polarity.

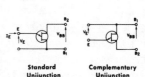

Standard Unijunction   Complementary Unijunction

We have chosen to use this polarity so that standard NPN planar passivated transistor processing techniques can be used. This results in a unijunction having superior stability and better uniformity than any unijunction previously available. The much tighter spread of intrinsic-standoff ratio now available is a significant advantage. For most applications, the polarity is not important.

WHAT CAN THE D5K DO?

The General Electric D5K can be used in most applications now using standard type unijunctions. Its unique stability and uniform properties make it ideal for stable oscillators, timers, and frequency dividers.

The key advantage of the D5K over conventional UJT's is its predictability over the specified temperature range. This allows an engineer to use design curves to select the correct R_{B2} compensating resistor instead of having to perform expensive temperature testing on individual devices.

For most applications now using conventional UJT's, the entire D5K2 population can be compensated in a given circuit with one resistor value by selecting the proper R_{B2} compensating resistor from Figure 2. For even better stability, a designer only has to measure the R_{BB0} of a device at room temperature, determine the proper R_{BB0}/R_{B2} ratio from Figure 3, and insert the correct R_{B2}. Using this method, oscillators and timers can be built offering 1.0% stability over most temperature ranges used.

Frequency dividers can be built with larger countdown ratios and drastically lower capacitor sizes due to the stability and low charge to trigger value (Q_t). Another product advantage, low base 1 to emitter voltage drop at high current, allows generation of high output pulses with *low* base to base voltages.

For further application information, refer to Application Note 90.72.

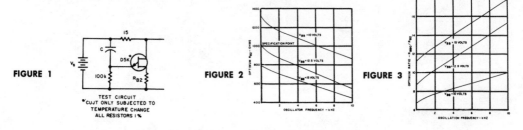

Fig. 4-17. Specifications for a typical unijunction transistor (courtesy of General Electric Co.). (Continued through page 117.)

D5K2

absolute maximum ratings: (25° C free air)

	D5K2	
Voltage		
Interbase Voltage	20	V
Current (Note 2)		
Average Emitter (Forward)	150	mA
Peak Emitter (Forward) (Note 1)	2	A
Peak Reverse Emitter	15	mA
Power		
Average Total (Note 2)	200	mW
Temperature		
Operating	−55 to +100	°C
Storage	−55 to +150	°C

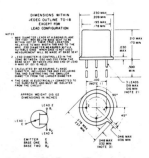

electrical characteristics: (25° C free air)

		Min.	Typ.	Max.	
Intrinsic Standoff Ratio (Note 3)	η	0.58	0.60	0.62	
Peak Point Voltage					
(V_{BB} = 5V)	V_P	3.2	3.45	3.7	Volts
(V_{BB} = 10V)	V_P	6.1	6.45	6.8	Volts
Interbase Resistance					
(I_{BB} = 0.1mA)	R_{BBO}	5	8	15	kohms
Emitter Breakdown Voltage					
(I_{EB1} = 10μA)	V_{EB1O}	8.0	9.5		Volts
Peak Point Current					
(V_{BB} = 10V)	I_P			15	μA
Valley Point Current					
(V_{BB} = 10V)	I_V	1	2		mA
Emitter Reverse Current					
(V_{EB1} = 5V)	I_{EB1O}		0.1	10	nA
Emitter Saturation Voltage					
(I_E = 50mA, V_{BB} = 10V)	$V_{E(sat)}$		1.1	1.5	Volts
Modulated Interbase Current					
(I_E = 50mA, V_{BB} = 10V)	$I_{B2(mod)}$		4	10	mA
Peak Pulse Voltage					
(Note 4)	V_{OUT}	3.5	4.5		Volts
Diode Voltage Drop					
(Note 3)	V_D	.30	.45	.60	Volts
Minimum Charge to Trigger					
(V_{BB} = 10V)	Q_t		50		pC
Turn-on Time (See Figure 7)	t_{on}			1	μsec.
Recovery Time (See Figure 7)	t_{rec}			10	μsec.
Relaxation Oscillator Frequency Shift from					
25°C Value (See Figure 1,					
C = 0.1μF, R_{B2} = 1kΩ, V_s = 12.5V)					
−15°C to +65°C			0.3	1.0	%
−55°C to +100°C			0.5	2.0	%

Notes:
1. For capacitor discharge, resistor current limiting is required for capacitors greater than 5 μF and recommended for all cases. (A minimum of 15 ohms is required for good temperature stability.)
2. Derate power and currents linearly to zero at maximum operating temperature.
3. The intrinsic-standoff ratio (η) is essentially constant with temperature and interbase voltage. It and the associated diode drop of peak point voltage are defined by the equations:

$$\eta = \frac{V_{P1} - V_{P2}}{V_{BB1} - V_{BB2}} \qquad V_D = V_{P2} - \eta V_{BB2} \qquad \text{Where: } V_{BB1} = 10V \pm .001V$$
$$V_{BB2} = 5V \pm .001V$$

4. The Base-One Peak Pulse Voltage is measured in the circuit shown in Figure 4. This specification is used to insure a minimum pulse amplitude for applications in SCR firing circuits and other types of firing circuits.

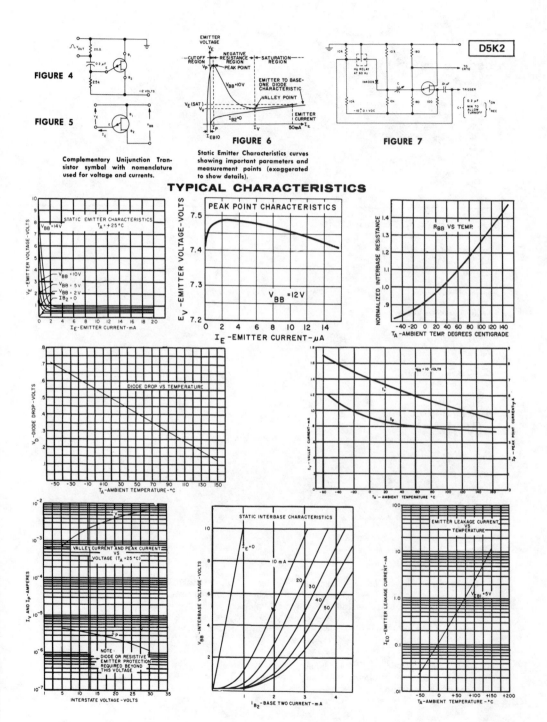

FIGURE 4

FIGURE 5

Complementary Unijunction Transistor symbol with nomenclature used for voltage and currents.

FIGURE 6

Static Emitter Characteristics curves showing important parameters and measurement points (exaggerated to show details).

FIGURE 7

TYPICAL CHARACTERISTICS

D5K2

APPLICATIONS

TYPICAL CIRCUITS

Since the CUJT has opposite polarities from standard UJT's, its oscillator circuit Figure (b) is "inverted." Figure (a) shows a positive, high energy pulse, while Figure (b) shows a negative. Circuit in Figure (c) results in positive pulses for SCR triggering.

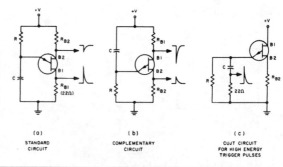

(a) STANDARD CIRCUIT
(b) COMPLEMENTARY CIRCUIT
(c) CUJT CIRCUIT FOR HIGH ENERGY TRIGGER PULSES

0.1 TO 90 SECOND TIMER

Timer interval starts when power is applied to circuit, terminates when voltage is applied to load. 2N2646 is used in oscillator which pulses base 2 of D5K. This reduces effective I_P of D5K and allows much larger timing resistor and smaller timing capacitor to be used than would otherwise be possible.

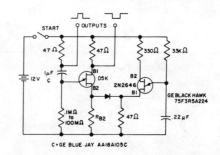

DECADE FREQUENCY DIVIDER

In next stage, product of R2 and C2 should be 10 × that of preceding stage (±2%). R2 should be between 27kΩ and 10 meg Ω.

C1 & C2—.0047 μF (±1%)
R1—100k Ω (±1%)
R2—1M Ω (±1%)
R3—R4—1k Ω (may need to be adjusted for variation of R_{BB} of CUJT)

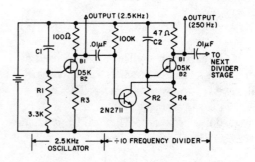

50 kHz OSCILLATOR

Higher frequency (stable) oscillators are now possible. Here are typical components for a 50 kHz circuit. This is possible because of the more nearly ideal characteristics of the D5K (over conventional UJT's). One application for higher frequency is TV horizontal oscillators. Note the low R_{B2}.

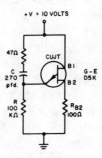

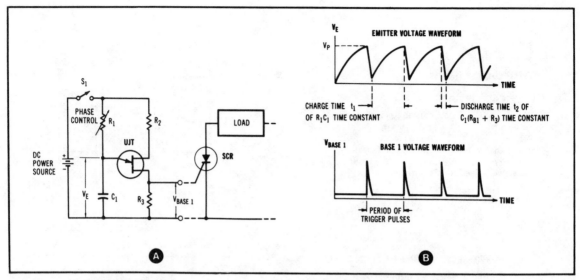

Fig. 4-18. A UJT relaxation oscillator circuit. (A) Basic circuit. (B) Voltage waveforms.

R_{BB} is the on-state minimum resistance of the UJT between base 1 and base 2, in ohms.

The value of $R_{BB(min)}$ is seldom provided in manufacturers' specifications. An approximation for $R_{BB(min)}$ is given by:

$$R_{BB(min)} = \eta\, R_{BB} \qquad \textbf{(Eq. 4-7)}$$

The UJT relaxation oscillator can be used as a trigger circuit for half-wave SCR, full-wave SCR, and full-wave triac power switching. This type of circuit offers the advantages for stable phase-delay triggering from approximately 0° to 180° and high peak trigger current for reliable turn-on of power thyristors.

The primary limitation of this UJT trigger circuit is the requirement for a dc power source and related single-polarity trigger signals. Since the UJT is a unilateral device with respect to current, the output trigger pulses are positive with respect to the negative terminal of the dc power source. This presents no unusual problems for triggering single SCR or triac switching circuits. Figure 4-19 shows a triac ac switching circuit employing a UJT relaxation-oscillator trigger. Note that dc power for operating the UJT is derived from the ac power source. The values of R1 and C1 are selected for the required triac firing angles. Resistors R2 and R3 are selected for the proper pulse amplitude for triggering the triac in the first and third quadrants. Resistor R4 is a current-limiting resistor for the UJT circuit.

Single-polarity trigger pulses, however, are *not* suitable for triggering two-SCR full-wave switching circuits. The two parallel SCRs illustrated in Fig. 4-20 require positive and negative trigger pulses with respect to the triggering circuit. A multiwinding pulse transformer is used in the UJT base-1 circuit to accomplish this requirement for providing both positive and negative trigger pulses.

The derived dc power sources in the UJT trigger circuits of Figs. 4-19 and 4-20 have two important advantages. First, the need for a separate and costly dc power supply is eliminated. The second advantage is probably more important; the varying dc voltage from the derived power source is *in phase* with the ac power source. This synchronizes the power thyristor turn-on with each ac cycle or half-cycle and ensures that the timing capacitor is discharged at the end of each firing interval.

PROGRAMMABLE
UNIJUNCTION TRANSISTOR

The *programmable unijunction transistor* (PUT)

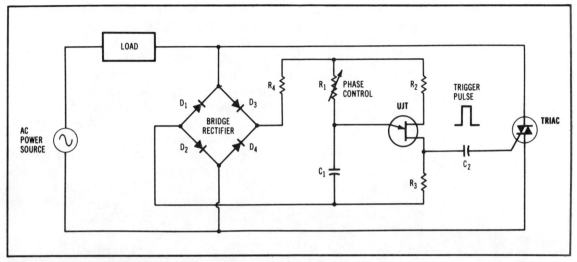

Fig. 4-19. A UJT phase-delay trigger circuit for triac full-wave switching.

is similar in operational characteristics to the UJT. The PUTs three terminals are designated as anode, gate, and cathode. This recent addition to the thyristor family looks and acts like an improved UJT. However, a microscopic examination of the internal structure of the PUT shows it to be another four-layer pnpn silicon device. Figure 4-21 provides some of these details.

The major advantage of the PUT over the UJT is the capability of programming key operating parameters with external resistors. A voltage-divider network connected to the gate terminal will provide control of the peak value (V_P) of the output signal.

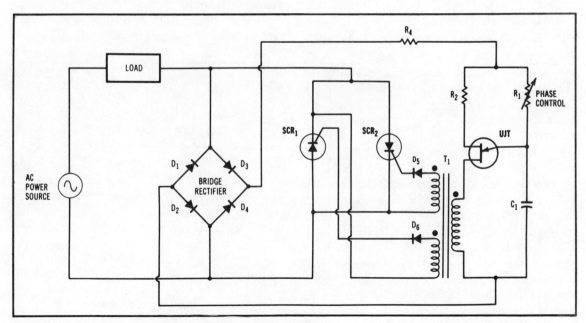

Fig. 4-20. A UJT phase-delay trigger for SCR full-wave switching.

119

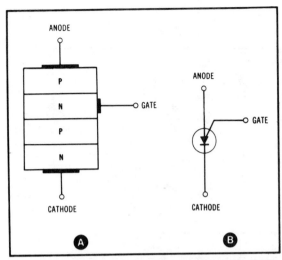

Fig. 4-21. The programmable unijunction transistor (PUT). (A) Basic construction. (B) Symbol.

Theory of Operation

Figure 4-22A illustrates the operation of the PUT in a test circuit. Voltage divider network R2-R3 provides a fixed gate voltage (V_G) to the gate terminal. Potentiometer R1 is connected as a variable voltage divider to control the anode voltage.

When the anode voltage is less than the gate voltage, the PUT is in a state of nonconduction, and only a negligible reverse leakage current flows through the PUT. This is due to the anode-gate pn junction being reverse biased. The VI characteristic curve in Fig. 4-22B shows this leakage current as I_{GAO}. When the anode voltage is raised above the gate voltage by approximately 0.7 V, the anode-gate pn junction is forward biased and permits turn-on of I_A. This value of anode voltage, called the peak voltage, is identified as V_P on the VI curve. At this point, the PUT latches into full turn-on and exhibits negative-resistance characteristics over a portion of the VI curve. By adjusting the ratio of the gate voltage-divider network, we can program the peak voltage (V_P) for different values. This also allows us to control the value of I_P.

The output from a PUT circuit is normally taken at the cathode terminal. In the off-state condition, V_{OUT} is nearly zero. When the PUT is triggered into conduction, V_{OUT} suddenly rises to a positive value controlled by the circuit design. When used as a thyristor trigger, V_{OUT} is adjusted to 2-6 V. The turn-on time is extremely fast—it can be as low as 80 ns.

Circuit design data for PUTs is usually included in manufacturers' specifications or data sheets. For example, the peak current (I_P) and peak voltage (V_P) values are specified in relation to equivalent gate voltage (V_S) and gate resistance (R_G). You can

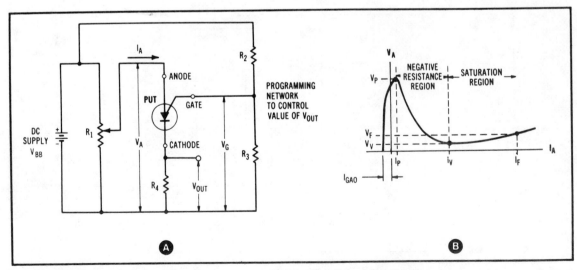

Fig. 4-22. Operating characteristics of a PUT. (A) Test circuit. (B) Anode characteristic curve.

calculate V_S and R_G for the circuit in Fig. 4-22 by the following equations. You may recognize these equations as the equivalent Thevenin voltage and resistance from dc circuit analysis.

$$R_G = \frac{R2\ R3}{R2 + R3} \quad \text{(Eq. 4-8)}$$

$$V_S = \frac{V_{BB}\ R3}{R2 + R3} \quad \text{(Eq. 4-9)}$$

PUT Specifications

Figure 4-23 provides specifications for a typical PUT, the D13T Series produced by the General Electric Co These PUTs are 100-volt versions of the 2N6027 and 2N6028 programmable unijunction transistors.

Applications

Like the UJT, the PUT is used in applications where a high-speed, low-power electronic switch is required. These applications include relaxation oscillators, timers, and phase-control circuits. The relaxation oscillator is useful for triggering high-power SCRs, since it is capable of producing pulsed high-current trigger signals.

Figure 4-24 shows a typical PUT relaxation-oscillator circuit and the pulsed output waveform at V_{OUT}. Note that a conventional UJT could be substituted in this circuit with resulting similar performance. The terminals of the PUT are equivalent to the UJT terminals as follows:

PUT	UJT
Anode	emitter
Cathode	Base 1
Gate	Base 2.

SILICON UNILATERAL AND BILATERAL SWITCHES

The *silicon unilateral switch* (SUS) and *silicon bilateral switch* (SBS) are among the newest and most advanced members of the thyristor family. These three-terminal devices are actually small *integrated circuits* (IC) containing transistors, zener diodes, and resistors. This type of construction provides improved performance as well as reduced costs.

The SUS and SBS possess negative-resistance switching characteristics similar to those of three- and four-layer diodes and unijunction transistors. They are capable of producing fast-rising, high-current trigger signals for power thyristors such as SCRs and triacs. Maximum power-dissipation ratings for the SUS and SBS are in the range of 300 mW to 350 mW.

Theory of Operation

Let us examine the theory behind the SUS first. The equivalent electrical circuit and symbol for this "circuit on a chip" are given in Fig. 4-25. This device is similar in performance to a low-power SCR, except that an anode gate is employed instead of the usual cathode gate.

The SUS is designed as a unilateral switching device—only current from anode to cathode should be allowed. Reverse current, if permitted, can result in damage to the SUS.

An npn and a pnp transistor are connected in a regenerative latch configuration. A zener diode (D1) is connected in parallel with the gate and cathode terminals. Current through the device is essentially zero as long as the applied voltage across the anode and cathode terminals is held to a value less than the rated switching voltage (V_S). When the applied voltage is increased to a value *equal* to V_S, the SUS is switched *on* due to the following conditions and actions:

A. The value of V_S must be equal to the emitter-base forward bias voltage of Q1 plus the zener breakdown voltage of zener diode D1. This is approximately 0.7 V plus 6.8 V for the circuit in Fig. 4-25. Manufacturers may rate this total of 7.5 V as 8 V to ensure reliable turn-on.

B. Current from the base of Q1 turns *on* the pnp transistor. High-gain pnp transistors are usually employed in this circuit to ensure fast and reliable

Silicon Programmable Unijunction Transistor (PUT)

PNPN
D13T SERIES
D13T3
D13T4

The General Electric D13T3 and D13T4 are 100 volt versions of the popular 2N6027 and 2N6028 Programmable Unijunction Transistors (PUT). These devices offer the designer the additional advantage of using higher circuit voltages thus improving timing stability.

For PUT application information please refer to Application Note 90.70.

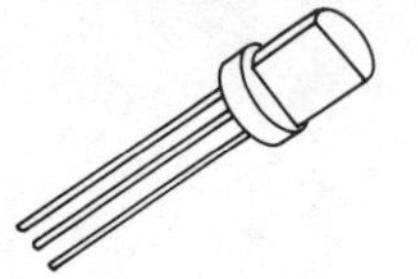

absolute maximum ratings: (25°C)

Voltage
Gate-Cathode Forward Voltage	+100 V
Gate-Cathode Reverse Voltage	—5 V
Gate-Anode Reverse Voltage	+100 V
Anode-Cathode Voltage	±100 V

Current
DC Anode Current†	150 mA
Peak Anode, Recurrent Forward	
(100μ sec pulse width, 1% duty cycle)	1 A
(20μ sec pulse width, 1% duty cycle)	2 A
Peak Anode, Non-recurrent Forward	
(10μ sec)	5 A
Gate Current	±20 mA

Capacitive Discharge Energy†† 250 μJ

Power
Total Average Power† 300 mW

Temperature
Operating Ambient† Temperature Range —50°C to +100°C

†Derate currents and powers 1%/°C above 25°C
††E = ½CV² capacitor discharge energy with no current limiting-non repetative

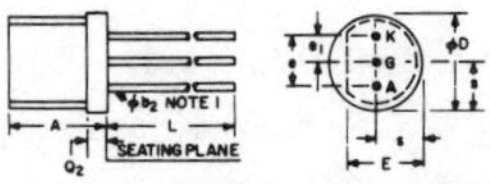

SYMBOL	INCHES		MILLIMETERS	
	MIN.	MAX.	MIN.	MAX.
A	.170	.265	4.32	6.73
φb₂	.016	.019	.406	.483
φD	.165	.205	4.19	5.21
E	.110	.155	2.79	3.94
e₁	.095	.105	2.41	2.67
e₁	.045	.055	1.14	1.40
L	.500		12.70	
Q₂		.075		1.90
	.080	.115	2.03	2.92

NOTE 1: LEAD DIAMETER IS CONTROLLED IN THE ZONE BETWEEN .070 AND .250 FROM THE SEATING PLANE. BETWEEN .250 AND END OF LEAD A MAX OF .021 IS HELD.

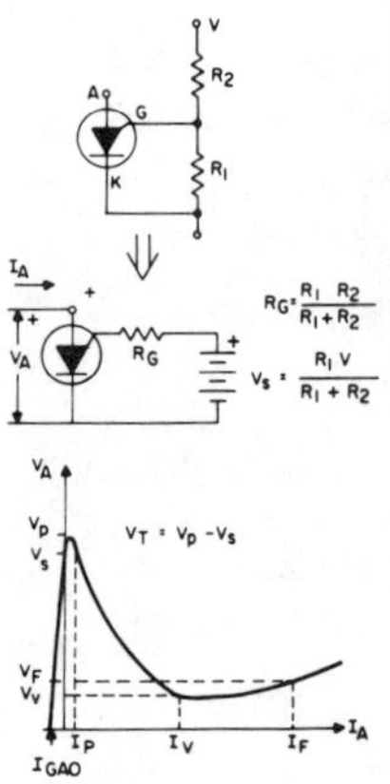

$V_T = V_P - V_S$

Figure 1
Figure 2 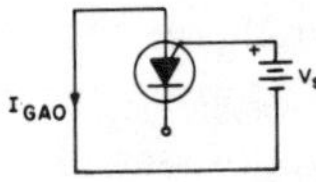

electrical characteristics: (25°C) (unless otherwise specified)

			D13T3		D13T4	
		Fig. No.	Min.	Max.	Min.	Max.
Peak Current	I_P	1				
(V_s = 10 Volts)						
(R_G = 1 Meg)				2		.15 μA
(R_G = 10 k)				5		1.0 μA
Offset Voltage	V_T	1				
(V_s = 10 Volts)						
(R_G = 1 Meg)			.2	1.6	.2	.6 Volts
(R_G = 10 k)			.2	.6	.2	.6 Volts
Valley Current	I_V	1				
(V_s = 10 Volts)						
(R_G = 1 Meg)				50		25 μA
(R_G = 10 k)			70		25	μA
(R_G = 200 Ω)			1.5		1.0	mA
Anode Gate-Anode Leakage Current	I_{GAO}	2				
(V_s = 100 Volts, T = 25°C)				10		10 nA
(T = 75°C)				100		100 nA
Gate to Cathode Leakage Current	I_{GKS}	3				
(V_s = 100 Volts, Anode-cathode short)				100		100 nA
Forward Voltage (I_F = 50 mA)	V_F			1.5		1.5 Volts
Pulse Output Voltage	V_O	4	6		6	Volts
Pulse Voltage Rate of Rise	t_r	4		80		80 nsecs.

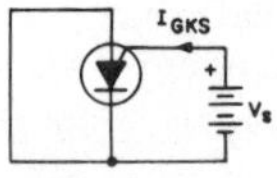

1069

Figure 3
Figure 4 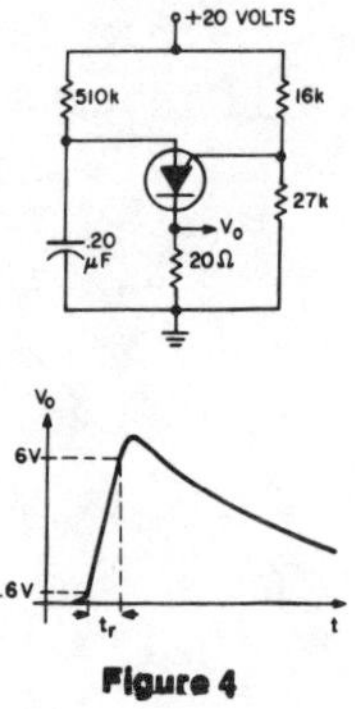

Fig. 4-23. Specifications for a typical programmable unijunction transistor (courtesy of General Electric Co.). (Continued through page 123.)

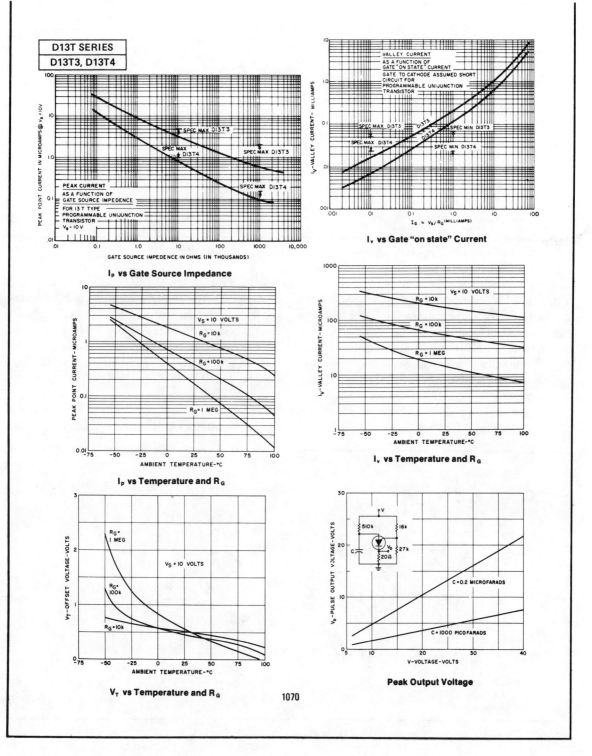

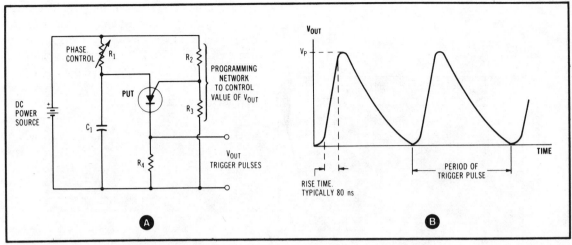

Fig. 4-24. A PUT relaxation oscillator. (A) Circuit diagram. (B) Output voltage waveform.

turn-on characteristics. Turn-on time for the SUS is typically on the order of 1 μs.

C. Collector current from Q1 flows into the base terminal of Q2 and collector resistor R1. The current into Q2 turns on the npn transistor.

D. The collector current of each transistor provides base current for the other transistor, and this regenerative action quickly drives *both* transistors into saturation. Typical SUS VI characteristics are illustrated in Fig. 4-26.

Once turned on, the SUS will remain in conduction until the anode current is reduced below the level of the holding current. Turn-off methods used in SCR circuits are also applicable to the SUS. As with the SCR, SUS turn-off time is longer than turn-on time. Depending on external circuit characteristics, SUS turn-off time may range up to about 25 μs.

The gate terminal may be used to turn on the SUS when the applied forward voltage is below

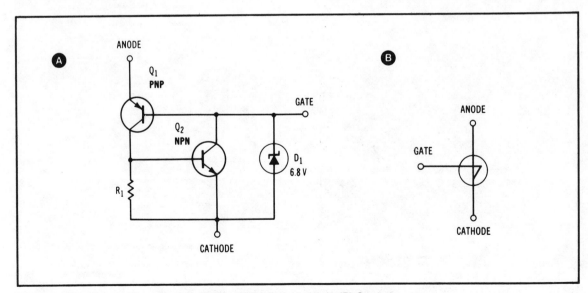

Fig. 4-25. The silicon unilateral switch (SUS). (A) Equivalent circuit. (B) Symbol.

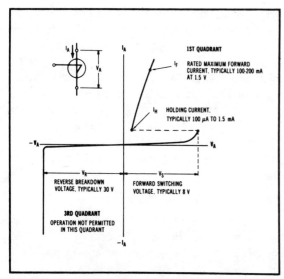

Fig. 4-26. Typical SUS VI characteristics.

V_S. This can be accomplished by switching a dc path between the gate and cathode terminals. Also, the gate terminal can be used in modifying the SUS turn-on characteristics. For example, an external zener diode with a lower breakdown voltage can be connected across the value of V_S required to turn on the device.

Now that we have covered the theory for the SUS, the operation of the silicon bilateral switch (SBS) is easy to understand. The SBS simply consists of two SUS circuits installed on the same IC chip and connected for bilateral current flow. Figure 4-27 gives this circuit configuration and the commonly used symbol.

Transistors Q1 and Q2 are connected as one of the two transistor latching circuits on the chip. With anode 1 positive, diode DI is forward biased and acts as a conventional diode rectifier. As the applied

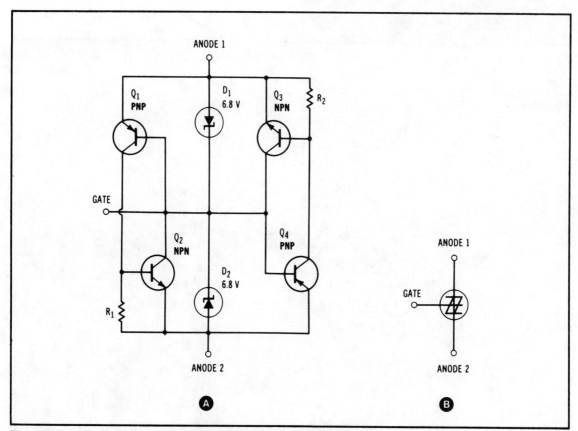

Fig. 4-27. The silicon bilateral switch (SBS). (A) Equivalent circuit. (B) Symbol.

voltage across the anode-1—anode-2 terminals is increased to V_S (the rated switching voltage), diode D2 undergoes zener breakdown and turns on Q1. This action, in turn, causes Q2 to conduct, and both transistors are quickly switched into saturation. The primary conduction path for a positive voltage on anode 1 is through Q1 and Q2. Transistors Q3 and Q4 remain off during this interval.

For reverse current, operation within the SBS is a mirror image of operation for forward current. When the applied voltage to anode 2 is positive, for example during the negative half-cycle of an ac voltage, Q4 and Q3 form the primary conduction path. During this half-cycle, D2 acts as a forward-biased diode, and D1 undergoes zener breakdown when the applied voltage equals V_S.

Figure 4-28 illustrates typical SBS VI characteristics. For most SBS devices, V_S ranges from about 6 V to 10 V. Due to uniform IC construction techniques, the SBS exhibits excellent symmetrical operational characteristics. For example, variations in switching voltages for forward and reverse current are held to about 0.2 V to 0.5 V. This characteristic is referred to as the absolute switching voltage difference in SBS specifications.

Like the SUS, the SBS can be controlled or turned on by the gate terminal. The maximum gate current for the off-state condition is about 5 mA. Maximum gate current for the on-state condition is limited only by the power-dissipation rating of the device.

SUS and SBS Specifications

Figures 4-29 and 4-30 provide specifications for typical SUS and SBS devices, respectively. These specifications contain design and applications data for use in circuit design.

Applications

The SUS and SBS devices are designed for

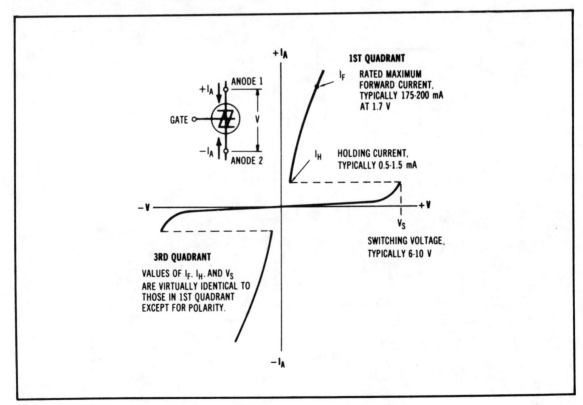

Fig. 4-28. Typical SBS VI characteristics.

Silicon Economy
Unilateral Switch
(SUS)

2N4987,90

The General Electric SUS is a silicon planar, monolithic integrated circuit having thyristor electrical characteristics closely approximating those of an "ideal" four layer diode. The device is designed to switch at 8 volts with a 0.02%/°C temperature coefficient. A gate lead is provided to eliminate rate effect, obtain triggering at lower voltages and to obtain transient free wave forms.

Silicon Unilateral Switches are specifically designed and characterized for use in monostable and bistable applications where low cost is of prime importance. These devices are in the low cost, TO-98 plastic package.

Applications Include:
- SCR Triggers
- Frequency Dividers
- Ring Counters
- Cross Point Switching
- Over-Voltage Sensors

EQUIVALENT CIRCUIT

CIRCUIT SYMBOL

DIMENSIONS WITHIN JEDEC OUTLINE TO-98

NOTE 1: Lead diameter is controlled in the zone between .070 and .250 from the seating plane. Between .250 and end of lead a max. of .021 is held.

ALL DIMEN. IN INCHES AND ARE REFERENCE UNLESS TOLERANCED

absolute maximum ratings:
(25°C free air) (unless otherwise specified)

Storage Temperature Range	−65 to +150	°C
Junction Temperature Range	−55 to +125	°C
Power Dissipation*	300	mW
Peak Reverse Voltage	−30	Volts
DC Forward Anode Current*	175	mA
DC Gate Current*†	5	mA
Peak Recurrent Forward Current (1% duty cycle, 10 μsec pulse width, $T_A = 100°C$)	1.0	Amp
Peak Non-Recurrent Forward Current (10 μsec pulse width, $T_A = 25°C$)	5.0	Amps

*Derate linearly to zero at 125°C.
†This rating applicable only in OFF state.
  Maximum gate current in conducting state limited by maximum power rating.

electrical characteristics: (25°C) (unless otherwise specified)

STATIC		2N4987			2N4990			
		Min.	Typ.	Max.	Min.	Typ.	Max.	
Forward Switching Voltage	V_S	6.0		10.0	7.0		9.0	Volts
Forward Switching Current	I_S			500			200	μA
Holding Current	I_H			1.5			.75	mA
Reverse Current								
($V_R = -30V$, $T_A = 25°C$)	I_R			0.1			0.1	μA
($V_R = -30V$, $T_A = 85°C$)	I_R			10.0			10.0	μA
Forward Current (off state)								
($V_F = 5V$, $T_A = 25°C$)	I_B			1.0			0.1	μA
($V_F = 5V$, $T_A = 85°C$)	I_B			10.0			10.0	μA
Forward Voltage Drop (on state)								
($I_F = 175$ mA)	V_F			1.5			1.5	Volts
Temperature Coefficient of Switching Voltage ($T_A = -55°C$ to $+85°C$)	T_C		±.02			±.02		%/°C
DYNAMIC								
Turn-on Time (See Circuit 1)	t_{on}			1.0			1.0	μsec
Turn-off Time (See Circuit 2)	t_{off}			25.0			25.0	μsec
Peak Pulse Voltage (See Circuit 3)	V_O	3.5			3.5			Volts
Capacitance (0V., f = 1 MHz)	C		2.5			2.5		pF

Fig. 4-29. Specifications for a typical silicon unilateral switch (courtesy of General Electric Co.). (Continued through page 130.)

2N4987, 90

PARAMETER DEFINITIONS

Static Characteristics

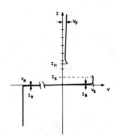

TEST CIRCUITS

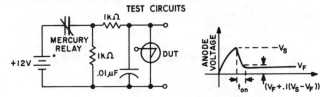

Circuit 1
Turn-on Time, t_{on}

Turn-on time is measured from the time the anode voltage first reaches V_s to the time where the anode voltage has fallen 90% of the difference between V_s and V_F.

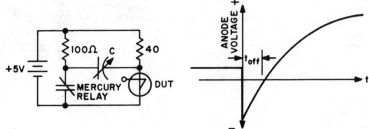

Circuit 2
Turn-off Time, t_{off}

The turn-off test is begun with the SUS in conduction and the relay contacts open. At $t = 0$ the contacts close and the anode is driven negative. C is adjusted downward, so that when the anode voltage becomes positive, the SUS just remains off. The turn-off time, t_{off}, is the time between initial contact closure and the point where the anode voltage passes up through zero volts. The capacitor is allowed to fully charge to 5 volts, at which time the contacts are reopened and the SUS triggers on.

Circuit 3
V_o

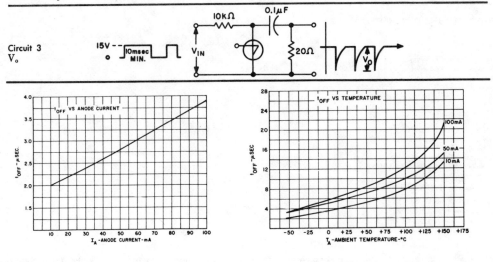

TYPICAL CHARACTERISTICS

2N4987, 90

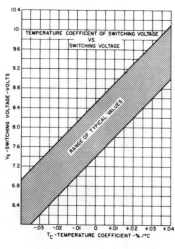
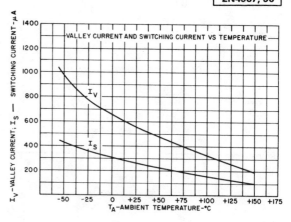
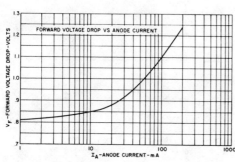
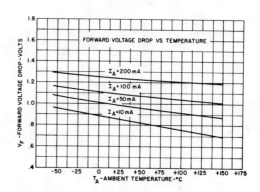
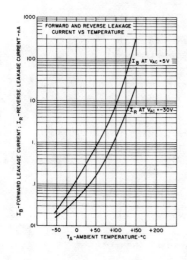
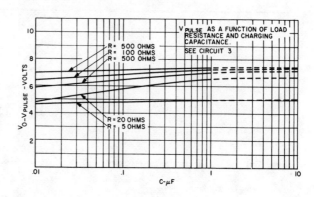

2N4987, 90

APPLICATIONS

BINARY DIVIDER CHAIN

Uses fewer components than transistor flip flops. Output at "B" gives transient free waveform.

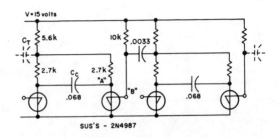

MOTOR SPEED CONTROL

Switching action of the 2N4990 allows smaller capacitors to be used while achieving reliable thyristor triggering.

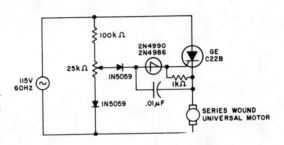

PULSE SHARPENERS

SUS is used to generate a rapid rise or fall time by using energy stored in a capacitor.

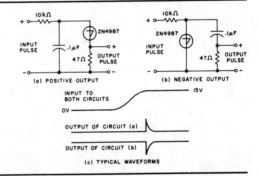

Ring Counter for Incandescent Lamps.

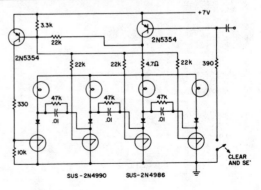

Silicon Economy
Bilateral Switch
(SBS)

2N4991

The General Electric SBS is a silicon planar, monolithic integrated circuit having the electrical characteristics of a bilateral thyristor. The device is designed to switch at 8 volts with a 0.02%/°C temperature coefficient and excellently matched characteristics in both directions. A gate lead is provided to eliminate rate effect and to obtain triggering at lower voltages.

The Silicon Bilateral Switches are specifically designed and characterized for applications where stability of switching voltage over a wide temperature range and well matched bilateral characteristics are an asset. They are ideally suited for half wave and full wave triggering in low voltage SCR and Triac phase control circuits. The 2N4991 is in the low cost, TO-98 plastic package.

absolute maximum ratings: (25°C free air) (unless otherwise specified)

Storage Temperature Range	−65 to +150	°C
Operating Junction Temperature Range	−55 to +125	°C
Power Dissipation*	300	mW
DC Forward Anode Current*	175	mA
DC Gate Current *†	5	mA
Peak Recurrent Forward Current (1% duty cycle, 10 μsec pulse width, $T_A = 100°C$)	1.0	Amp
Peak Non-Recurrent Forward Current (10 μsec pulse width, $T_A = 25°C$)	5.0	Amps

*Derate linearly to zero at 125°C.
†This rating applicable only on OFF state. Maximum gate current in conducting state limited by maximum power rating.

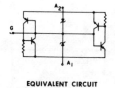

EQUIVALENT CIRCUIT

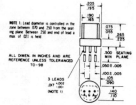

CIRCUIT SYMBOL

electrical characteristics:** (25°C) (unless otherwise specified)

		Min.	Typ.	Max.	
STATIC					
Switching Voltage	V_S	6		10	V
Switching Current	I_S			500	μA
Absolute Switching Voltage Difference	$\|V_{S2} - V_{S1}\|$			.5	V
Absolute Switching Current Difference	$\|I_{S2} - I_{S1}\|$			100	μA
Holding Current	I_H			1.5	mA
Current (Off State)					
($V_F = 5V$, $T_A = 25°C$)	I_B			1.0	μA
($V_F = 5V$, $T_A = 85°C$)	I_B			10.0	μA
Temperature Coefficient of Switching Voltage ($T_A = -55°C$ to $+85°C$)	T_C		±.02		%/°C
Forward Voltage Drop (On State) ($I_F = 175$ mA)	V_F			1.70	V
DYNAMIC					
Turn-on Time (See Circuit 1)	t_{on}			1.0	μsec
Peak Pulse Amplitude (See Circuit 3)	V_o		3.5		V
Turn-off Time (See Circuit 2)	t_{off}			30.0	μsec

**This device is a symmetrical negative resistance diode. All electrical limits shown apply in either direction of current flow.

Fig. 4-30. Specifications for a typical silicon bilateral switch (courtesy of General Electric Co.). (Continued through page 134.)

2N4991

STATIC CHARACTERISTICS

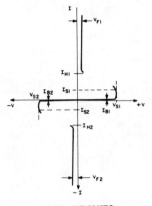

TEST CIRCUITS

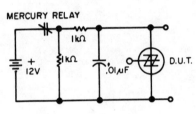

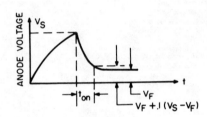

Circuit 1
Turn-on Time, t_{on}

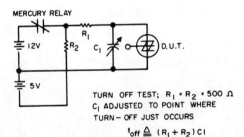

TURN OFF TEST; $R_1 = R_2 = 500\,\Omega$
C_1 ADJUSTED TO POINT WHERE
TURN-OFF JUST OCCURS

Circuit 2
Turn-off Time, t_{off}

$t_{off} \triangleq (R_1 + R_2) C_1$

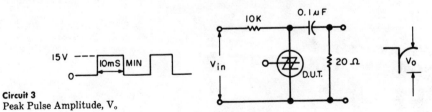

Circuit 3
Peak Pulse Amplitude, V_o

TYPICAL CHARACTERISTICS

2N4991

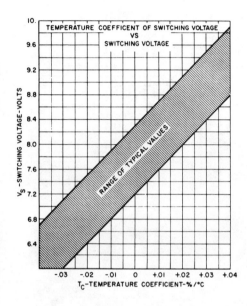

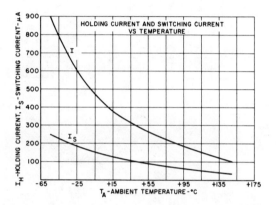

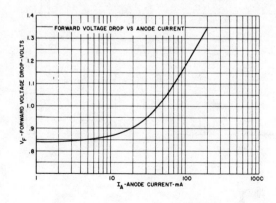

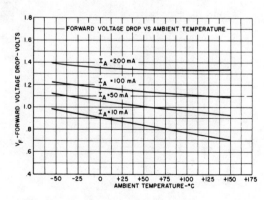

2N4991

TYPICAL CHARACTERISTICS

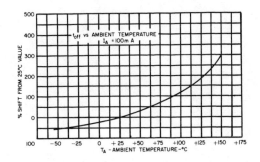
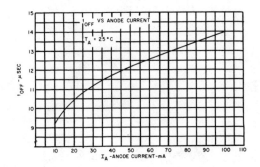
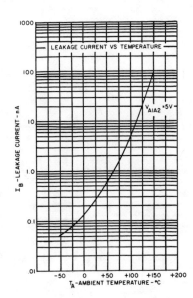
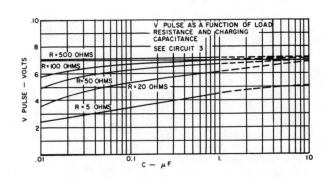

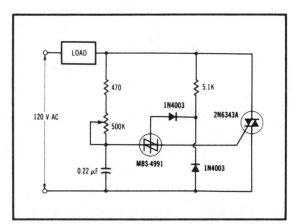

Fig. 4-31. Full range lamp dimmer (courtesy of Motorola Semiconductor Products Inc.).

high-speed signal switching applications where switching voltage stability and low cost are required. In addition to serving as triggers for power thyristors, the SUS and SBS are employed in digital circuits involving frequency dividers, ring counters, bistable memory circuits, and pulse generators. Other related applications include voltage sensing in such circuits as the "electronic crowbar," or overvoltage protection for dc power supplies.

One interesting circuit using an SBS trigger is the full-range lamp dimmer shown in Fig. 4-31. This is referred to as a *hysteresis-free power controller*. The gate teminal is employed to eliminate "flash-on" of the lamp when the phase control is adjusted for no power to the load. Without the gate terminal connection, the timing capacitor cannot discharge through the SBS during each ac cycle. Therefore, the voltage across the capacitor can build up and eventually trigger the SBS; this results in triac turn-on and a momentary flash of the lamp. The 1N4003 silicon rectifiers permit flow of gate current at the end of each ac positive half-cycle, turning on the SBS. The triac is triggered, and the timing capacitor discharges to near zero volts. With a proper heat sink, the 2N6343A triac is capable of handling load current up to about 12 A.

Chapter 5

Planning the Thyristor Control System

THYRISTOR CONTROL SYSTEM DESIGN INvolves many factors. Due to space limitations, this broad area of the thyristor field cannot be covered completely in this book. Many excellent design handbooks are available to the designer. We will cover some of the most important areas that should be considered in designing thyristor control circuits.

An important first step in thyristor circuit design is defining the user's requirements. The documentation of these basic operational requirements helps to ensure construction of a usable product and to avoid subsequent costly modifications.

INITIAL DESIGN

A preliminary analysis of the user's requirements will identify basic design parameters. Load characteristics in terms of voltage, current, power, and power factor will determine the power-handling capability and other maximum electrical ratings for the thyristor. Gate triggering modes, such as phase control or zero-voltage switching, will further help to narrow the selection of power thyristors for the intended application.

The trigger-circuit design will depend on the type of control required for the intended application as well as gate-trigger signal characteristics. An externally controlled trigger circuit, either automatic or semiautomatic, will necessitate some form of interface with the control system. Fail-safe and emergency backup operating modes may be required for some installations.

ENVIRONMENTAL CONSIDERATIONS

Many thyristors used in power switching systems are installed in "hostile" operating environments. A detailed knowledge of the operating conditions is necessary to ensure that the thyristor circuit design will result in a reliable and economical system.

The major factors that should be considered in designing a new thyristor control system are listed below. These factors are also applicable to the maintenance of existing installations, especially when deficiencies or modifications are being investigated.

1. Maximum power loads to be switched including possible overvoltage or overcurrent conditions.
2. Minimum and maximum ambient temperatures.
3. Adequate cooling and possible heat-sink requirements.
4. Proper mechanical installation.
5. Protection against power and signal line-voltage transients.
6. Elimination or suppression of radio-frequency interference (rfi).

These factors are not given in any order of priority. The disregard of any of the above factors can lead to poor reliability and possible catastrophic failure of the power thyristors and other circuit components.

The failure of other components in the power thyristor switching circuit or the associated load circuit may produce a sudden inrush of current and destroy the power thyristor. Even worse, thyristor control circuit malfunctions may lead to damage of related load equipment or down-time of assembly-line production.

Once the thyristor operating environment has been identified, final selection of specific power thyristors and thyristor circuit design can be accomplished. Let us examine some of the factors involved in selecting thyristors for a particular application.

SELECTING THE PROPER THYRISTOR

The initial analysis of the thyristor power switching requirements will identify maximum power loads and the related voltage and current levels. It is a good idea to thoroughly investigate the actual operating environment and determine if overvoltage or overcurrent conditions will cause problems. Voltage transients from nearby electrical equipment such as arc welders may result in premature firing of SCRs and triacs. A more serious possibility is catastrophic failure of power thyristors due to reverse voltage breakdown.

The key to selecting the proper thyristor for a given task is the manufacturers' specification sheets. Maximum device ratings for each thyristor are given in terms of voltage, current, power, and operating temperature levels. These basic parameters will show whether the device selected will perform and survive in the operating environment. If any of these maximum ratings is exceeded, catastrophic failure can result.

Manufacturers employ varying terminology in defining thyristor specifications. Sometimes the confusion over specific device ratings leads to a communications gap between the user or designer and the manufacturer. For example, a maximum gate trigger current (I_{GT}) for a certain SCR might be assumed to be the upper limit in terms of gate power dissipation. In actuality, the manufacturer may be attempting to state that this is the minimum gate trigger current for any member of a given series of devices. In many instances, experienced users and designers have learned to interpret manufacturers' specifications by trial-and-error methods.

The major thyristor manufacturers usually provide design handbooks or application notes for assisting designers of thyristor circuits. You can find these types of publications at electronics parts houses or you can contact the manufacturers directly. One manufacturer has developed a series of checklists for identifying specific device requirements. Figure 5-1 shows a copy of the phase-control SCR application checklist.

Also, virtually all manufacturers of electrical and electronic components and devices cooperate in standardization efforts to ensure interchangeability. For example, case types and sizes, device rating parameters, and uniformity in acceptance testing contribute to multiple source access.

As you may have observed, thyristors and most other semiconductor devices are assigned either 1N, 2N, or 3N designators, or a commercial designator. The N numbers are registered with the Electronic Industries Association (EIA) and assure compatibility between different manufacturers.

Commercial standardization of thyristors and other semiconductor devices is handled by the Joint

ⓦ FAST SWITCHING SCR APPLICATION CHECKLIST
(Make copy for each use)

1. APPLICATION: _____
_____

2. CIRCUIT:
Sketch circuit (showing all component values including inductances) or attach drawing

3. CIRCUIT VOLTAGE:
- Peak forward blocking voltage, V_{DRM} _____ V
- Peak reverse blocking voltage, V_{RRM} _____ V
- Maximum expected transient voltage _____ V
- Desired voltage safety factor _____
- Preferred SCR voltage rating _____ V
- Other _____

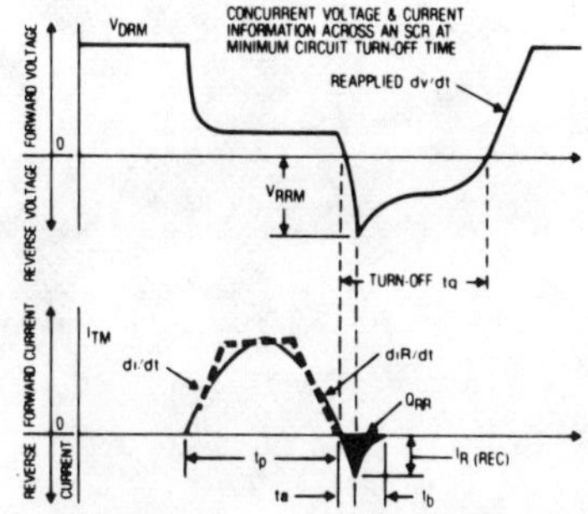

CONCURRENT VOLTAGE & CURRENT INFORMATION ACROSS AN SCR AT MINIMUM CIRCUIT TURN-OFF TIME

4. CIRCUIT CURRENT*: (if required, sketch waveform(s) on back of page)

Worst Case Pulsewidth Conditions
- Maximum peak current, I_{TM} _____ A
  - Pulsewidth, t_p _____ μs
  - Waveshape
- Initial rate of current rise, di/dt _____ A/μs
- Rate of current fall, di_R/dt _____ A/μs
- Reverse recovered charge, Q_{RR} _____ μcoul.
  - Max. overshoot current $I_{R(REC)}$ _____ A
  - Recovery time, t_a _____ μs
  - Recovery time t_b _____ μs
- Peak surge current (waveform) _____ A
  - Pulsewidth _____
  - Number of cycles _____
  - Assymetry (L/R ratio) _____
- Other _____

*If paralleling is required, state method of current sharing: _____

5. SWITCHING:
- ☐ Soft Commutation or ☐ Hard Commutation
- Maximum available turn-off time, t_q ____ μs @ T_J ____ °C
- Reapplied dv/dt _____ V/μs
- Maximum operating frequency _____ or duty cycle _____ %

6. GATE DRIVE AVAILABLE:
- I_{G2} _____ A • t_{GP} _____ μs
- t_{GR} _____ μs • t_{GS} _____ μs
- $I_{GT (max)}$ @ 25 °C _____ ma
- Forward gate source voltage _____ V

7. THERMAL:
- Cooling Medium (check one) -
  - ☐ AIR - ☐ Natural Convection, altitude _____ feet or
  - ☐ Forced _____ LFM _____ CFM**

** Duct cross-sectional area _____

  - ☐ WATER - _____ GPM flow rate
  - ☐ OIL (immersed) - Type _____ Manufacturer _____
  - ☐ OTHER - _____
- Cooling medium maximum temperature, °C _____
- If heat sink is known, specify $R_{\theta SA}$ _____ °C/W.
- Other thermal considerations _____

8. MECHANICAL:
- Desired package type (check one) -
  - ☐ STUD MOUNT
  - ☐ DISC MOUNT
  - ☐ INTEGRAL HEAT SINK
  - ☐ FLAT BASE
  - ☐ OTHER _____
  - ☐ NO PREFERENCE
- Special size, weight, or other restrictions _____

9. PROJECT REQUIREMENTS: Quotation Due Date _____
- Quantity Required _____ Person Requesting Information
- Timetable _____ Name _____
- Long-range Potential _____ Phone _____ X _____
- Other Remarks _____ Job Function _____
  _____ Company _____
  _____ Address _____
- ☐ Special screening and/or high reliability test requirements are attached.
- ☐ Also quote on this application using ⓦ assemblies.

City _____ State _____
Bldg. _____ Zip _____

Please complete a copy of this form for each different application. Forward this form to Westinghouse Electric Corporation, Semiconductor Division, Attention: Sales Department, Youngwood, Pa. 15697 for complete quotation. If you need faster service, please call (412) 925-7272 for a quote.

Fig. 5-1. Designer's checklist for phase-control SCR circuits (courtesy of Westinghouse Electric Corp.).

Electron Device Engineering Council (JEDEC). This group is sponsored by the EIA and the National Electrical Manufacturers Association (NEMA). Military standards for electronic components are developed and controlled by the U. S. Department of Defense or subordinate military agencies.

THERMAL CONSIDERATIONS

The successful operation of any semiconductor device is a direct function of the thermal environment. More specifically, the junction temperature of the device controls how much power can be safely dissipated before thermal runaway or meltdown is experienced. Failure rates of semicon-

Fig. 5-2. Heat sinks for TO-5 cases (courtesy of Thermalloy Inc.).

ductor devices, including thyristors, are dramatically reduced when the junctions are maintained at low temperature levels.

As a general rule, low-power lead-mounted thyristors with power-dissipation ratings of less than one watt do not require heat sinks. These devices are usually soldered to printed circuit (pc) boards or terminal strips. They depend mostly on cooling by radiation and convection from the case to the surrounding air. Some conduction of heat is also accomplished through the leads.

Low-power thyristors are packaged in small cases such as the TO-5 metal can or TO-92 plastic case. Cooling fins (Fig. 5-2) can be employed when the device is used in high-temperature environments.

Junction temperatures of power thyristors can be held within safe operating limits by installing the devices on heat sinks. Sometimes referred to as a *heat exchanger* or *cooler*, the *heat sink* is a metallic base for conducting heat away from the thyristor and into the surrounding environment. If the thyristor is called upon to dissipate average power levels of over one watt, some form of heat sink will be required to maintain safe junction temperatures.

The Thermal Spectrum

Thermal specifications for thyristors contain operating temperature limits in terms of either junction or case temperature. A single temperature specification for storage conditions is employed, since both case and junction are maintained at the same temperature. Limits for typical power SCRs and triacs are as follows:

Operating Conditions:
   Junction Temperature (T_J): $-40°$ C to $+125°$ C
   Case Temperature (T_C): $-40°$ C to $+100°$ C
Storage Conditions:
   Storage Temperature (T_{stg}): $-40°$ C to $+150°$ C

Figure 5-3 shows a portion of the thermal spectrum and compares semiconductor operation and storage temperatures with other values of temperatures. Note that the theoretical upper limit for silicon devices is approximately 175° Celsius. This represents the junction temperature of the device. However, due to practical considerations, thyristor junctions must be held to lower temperatures. Specific operating temperature limits depend on the type of device being used. This is due to thermal-instability problems and a lowering of breakdown voltage ratings at the higher temperatures. (Notice that the Celsius scale is still often referred to by the name *centigrade* as in Fig. 5-3.)

Lower operating and storage temperature limits are due primarily to mechanical stress problems involved with the silicon pellet and related materials used in fabricating thyristors. The varying thermal coefficients of expansion of the different materials can cause the silicon structure to fracture or be dislodged from the bonded connections. The lower temperature limits of $-40°$ C noted above apply to commercial-grade thyristors. Some military electronics standards require lower operating and storage temperature limits on the order of $-65°$ C. Special and costly manufacturing techniques are required to produce these devices.

Heat Flow in Thyristors

The heat generated within a thyristor depends primarily upon the voltage drop across and the current through the device. Since most thyristors are operated close to their upper temperature limits, the junction temperatures of the silicon structure will range up to 100° C or more. This heat must be effectively dissipated—otherwise the thyristor will be destroyed. Small, low-power thyristors dissipate this heat directly into the surrounding air and through conduction in the leads. However, the ambient temperature must be sufficiently low to permit this flow of heat. For normal commercial standards, the ambient temperature surrounding the device should never be allowed to rise above about 55° C.

Power thyristors are fabricated with the silicon pellet mounted directly to the metal case housing.

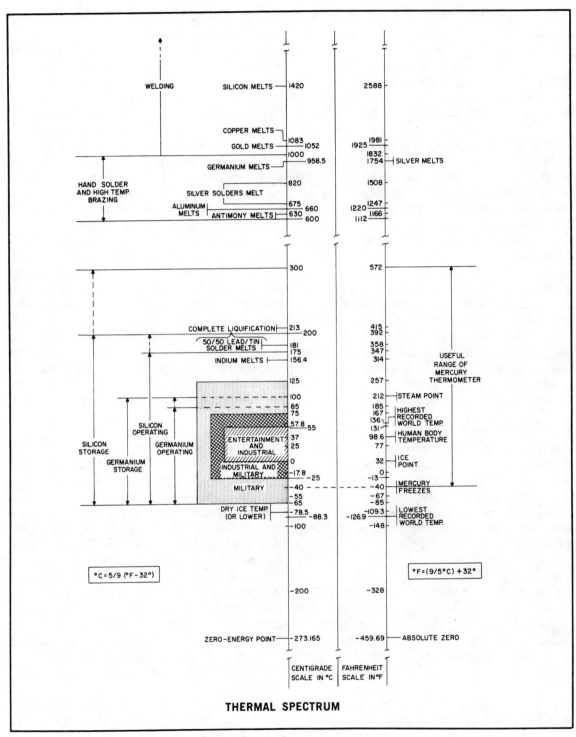

Fig. 5-3. Thermal spectrum (courtesy of General Electric Co.).

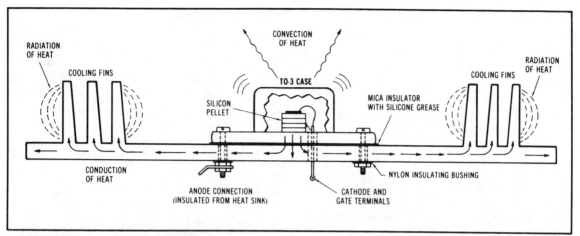

Fig. 5-4. Heat flow for a thyristor installed on a heat sink.

This provides a direct path for heat conduction from the silicon pellet to the metal case. Additional heat transfer is accomplished when the thyristor case is mounted on a heat sink.

Mounting the thyristor pellet directly to the case housing results in an electrical connection between the pellet and the metal case. Thus, the case serves as one of the main current terminals (such as the anode for an SCR or main terminal for a triac). The manufacturers' specification sheets will provide the specific electrical configuration and required connections.

Heat is transferred from the thyristor to the heat sink and surrounding ambient air by the processes of conduction, convection, and radiation. Figure 5-4 illustrates these three methods of heat transfer by means of a cross-sectional view of a thyristor installed on a heat sink. Note that the TO-3 case is installed with a mica insulating washer for electrical isolation between the case and the heat sink.

Conduction involves the direct transfer of heat from the junction to the case and from the case to the heat sink. This is the most effective method for transferring heat from the thyristor to the heat sink. Convection is the transfer of heat from the thyristor and heat sink to the surrounding ambient air. Air flow is a major factor in the transfer of large amounts of heat. Radiation is the process of emitting heat from an environment of higher temperature to an environment of lower temperature. The additional area created by the cooling fins provides for efficient radiation.

Thermal Conductivity

Physicists describe heat as a form of energy that can be measured in either *British Thermal Units* (BTU) or *calories*. Because of the present emphasis on the metric system, we will discuss heat flow relationships in terms of calories and Celsius (centigrade) temperature units. If desired, temperature values in Celsius can be converted to Fahrenheit by using the following relationship:

$$°F = \frac{9}{5} (°C) + 32 \qquad \textbf{(Eq. 5-1)}$$

where,
°F is the temperature in degrees Fahrenheit,
°C is the temperature in degrees Celsius.

The calorie is defined as the quantity of heat, or energy, necessary to raise the temperature of one gram of water from 15°C to 16°C. In electrical terms, one calorie is equivalent to 4.18 watt-seconds.

Heat generated by an energy source flows toward regions of lower temperature. In the conduction process, energy in the form of heat is passed from molecule to molecule. The rate of heat flow is proportional to the cross-sectional area of the conductor, the temperature gradient (the difference in temperature from source to sink per unit of length), and the type of material being used. The following equation expresses this relationship in mathematical terms:

$$H = kA \frac{(T_2 - T_1)}{d} \quad \text{(Eq. 5-2)}$$

where,
H is the rate of heat flow through the conductor in calories per second,
k is the coefficient of thermal conductivity for the conductor material,
A is the cross-sectional area,
$t_2 - t_1$ is the temperature gradient,
d is the length of the conductor.

For the metric system, A is expressed in square centimeters, d is in centimeters, and temperature is in degrees Celsius. Table 5-1 shows the thermal conductivity characteristics for some of the more common materials used in electronics. This table illustrates that metals conduct heat much more efficiently than materials such as glass, mica, or air.

Thermal Resistance

Equation 5-2 could be used to calculate heat flow characteristics for thyristors if the manufacturer provided specific details concerning the construction of the particular thyristor. Fortunately, the analysis of heat flow in the thyristor and heat sink is simplified by the thermal resistance rating for each thyristor and commercial heat sink. Only when custom heat sinks are involved does Equation 5-2 come in handy.

The thermal resistance of a material may be considered to be the opposition the material presents to the flow of heat. Thermal resistance is the inverse of thermal conductivity (similar to the concept that electrical conductivity is the inverse of electrical resistance).

Virtually all thyristor specifications provide thermal resistance data in terms of junction-to-case ratings, referred to as θ_{JC} or $R_{\theta JC}$. Typical values of θ_{JC} range from about 0.05° C/watt for high-current SCRs to about 25° C/watt for medium-power SCRs and triacs. Some specification sheets also contain thermal resistance ratings for case to heat sink (θ_{CS}) or case to ambient (θ_{CA}). The θ_{CA} rating is particularly useful for analyzing low-power thyristor operation where heat sinks may not be required.

The value of θ_{CS} for each thyristor installation will vary due to the differences in cases, heat sinks, and mounting methods. Handbooks or manufac-

Type of Material	Thermal Conductivity* (Calories per second per square centimeter)
Aluminum	0.504
Air	0.000054
Copper	0.918
Glass	0.0015
Iron	0.161
Ice	0.00396
Lead	0.083
Mica	0.0018
Silicon	0.201
Silver	0.974
Water	0.00143
Zinc	0.265

*The specific values given do not reflect a common test temperature and are subject to minor variations when compared to different test temperatures.

Table 5-1. Thermal Conductivity of Various Common Materials.

turers' application notes contain technical data and nomographs for calculating case-to-heat-sink thermal resistance. Factors to be considered include the use of thermal compounds, electrical isolation, and clamping force between the thyristor and heat sink.

Thermal Compounds

Ideally, the mounting surfaces between the thyristor and heat sink should be perfectly flat for maximum heat transfer. Although manufacturing techniques can produce surfaces that are flat within 0.0005 to 0.002 inch, minute surface irregularities contain pockets of air which reduce heat transfer. Furthermore, continued expansion and contraction of the metal surfaces and corrosion can aggravate this condition.

A number of thermal conducting compounds, such as silicone grease, have been developed to improve heat transfer characteristics. Also, the sealing action of a thermal compound helps to prevent corrosion of the mounting surfaces. The use of a thermal compound or grease can reduce the case-to-heat-sink thermal resistance by a factor of 1.5 to 2 or more. For example, the θ_{cs} for a TO-220 case mounted to a heat sink can be reduced from about 5.25° C/watt for a dry mounting to about 2.0° C/watt if a thermal compound is used. Table 5-2 provides a list of some of the commercial compounds available for thyristor installations.

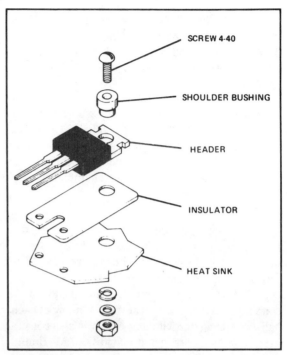

Fig. 5-5. Typical mounting arrangement for a TO-220 case (courtesy of RCA).

Table 5-2. Typical Thermal Compounds Used in Thyristor and Other Semiconductor Installations.

Thermal Compound	Manufacturer
No. 340 Silicone Heat Sink Compound	Dow Corning Company
No. G623 Silicone Grease	General Electric Company
Thermate	International Electronic Research Corporation
Thermalcote Thermal Joint Compound	Thermalloy, Inc.
Type 120 Thermal Joint Compound	Wakefield Engineering, Inc.

Electrical Isolation

The metal case for almost all thyristors serves as one of the electrical connections to the silicon pellet. This type of construction allows maximum heat transfer from the pellet to the case. Figure 5-4 illustrates this concept.

In some applications, the thyristor case must be *electrically isolated* from the heat sink or the circuit ground. This can be accomplished by a variety of methods. One solution is to insulate the heat sink from the main chassis with Teflon or nylon insulators and washers. Another method is to use thin mica or epoxy-coated aluminum wafers installed between the case and heat sink. This is a popular method for isolating medium-power thyristors contained in the TO-3, TO-220, or similar cases. Figure 5-5 shows a TO-220 installation layout with an insulating wafer.

Electrical isolation of the thyristor case does, however, cause the case-to-heat-sink thermal resistance to be increased. For example, a 0.003-inch mica wafer installed on a TO-220 case will increase the "dry" value of θ_{CS} from about 5 to 10° C/watt. Applying a silicone grease to both sides of the mica wafer reduces the value of θ_{cs} to about 3.75° C/watt.

A more recent development is the use of beryllium oxide insulating wafers, which exhibit a thermal conductivity comparable to that of aluminum. Some heat sinks are available with special coatings that provide both electrical isolation and excellent heat transfer characteristics.

Heat-Sink Analysis

Current conduction in a thyristor may be described in terms of *steady-state*, *repetitive-pulse*, or *nonrepetitive-pulse* operation. For example, an SCR turned on in a dc control circuit for a substantial period of time is considered to be in a steady-state condition. A triac gated on at the beginning of each half-cycle in an ac switching circuit is also considered to be in a steady-state condition. Maximum power dissipation occurs during steady-state operation.

Repetitive-pulse operation for thyristors in either ac or dc control circuits involves a repetitive, or cyclic, turn-on and turn-off of current within short periods of time. For example, an SCR used in the horizontal-deflection circuit of a television receiver or the capacitive-discharge ignition system of an automobile is operating in a repetitive-pulse state. In this type of circuit, we are concerned with average power dissipation.

Nonrepetitive-pulse, or single-pulse, operation, as the term implies is simply turning the thyristor on for a short period of time. For example, an SCR may be turned on for a single 1-μs period. Peak power dissipation may be extremely high while average power dissipation is very low in this type of operation. Fast-switching SCRs are representative of thyristors used in both repetitive- and nonrepetitive-pulse switching circuits.

Thermal resistance ratings in thyristor specification sheets are given in terms of steady-state current conduction. Under these conditions, a uniform junction temperature is assumed along with a constant transfer of heat from the junction to the case. Thus, design based on thermal resistance ratings represents an overkill for repetitive- or nonrepetitive-pulse operation. Some specifications, particularly those for fast-switching SCRs, provide additional data for repetitive or momentary surge conditions.

Steady-State Conduction—The basic equation for thermal resistance as used in thyristor heat-sink analysis for steady-state heat flow is:

$$\theta_{21} = \frac{(T_2 - T_1)}{P_d} \quad \textbf{(Eq. 5-3)}$$

where,
θ_{21} is the thermal resistance in degrees Celsius per watt between points 2 and 1 along a heat-conducting system,
$(T_2 - T_1)$ is the temperature differential between the two points,
P_d is the power being dissipated by the semiconductor device in watts.

Figure 5-6 shows the equivalent heat circuit for the thermal resistances encountered in a typical power thyristor installed on a heat sink. Like electrical resistors in series, the total thermal resistance

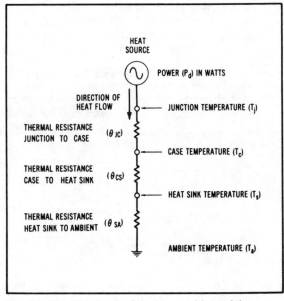

Fig. 5-6. Thermal circuit of thyristor and heat sink.

for this model is the sum of the individual thermal resistance values, or:

$$\theta_{JA} = \theta_{JC} + \theta_{CS} + \theta_{SA} \quad \text{(Eq. 5-4)}$$

where,
θ_{JA} is the total thermal resistance from the semiconductor junction to the ambient medium (such as the air),
θ_{JC} is the thermal resistance between the semiconductor junction and case,
θ_{CS} is the thermal resistance between the semiconductor case and the heat sink,
θ_{SA} is the thermal resistance between the heat sink and the ambient medium.

All thermal resistance values are expressed in degrees Celsius per watt. Thyristor and heat sink manufacturers provide this information in their specification sheets.

We can now revise the above equations to provide two practical equations for use in analyzing thyristor heat-sink installations. First, we can determine the maximum power dissipation for a given thyristor-heat-sink system at specified temperatures:

$$P_d = \frac{T_{j(max)} - T_a}{\theta_{JA}} \quad \text{(Eq. 5-5)}$$

or

$$P_d = \frac{T_{j(max)} - T_a}{\theta_{JC} + \theta_{CS} + \theta_{SA}} \quad \text{(Eq. 5-6)}$$

where,
P_d is the power dissipation in watts,
$T_{j(max)}$ is the maximum junction temperature specified by the manufacturer, in degrees Celsius,
T_a is the ambient temperature surrounding the thyristor and heat sink, in degrees Celsius,
θ_{JA}, θ_{JC}, θ_{CS}, and θ_{SA} are as defined above.

Equation 5-3 can be revised to solve for the maximum junction temperature when the maximum power dissipation and ambient temperature values are known. This relationship is:

$$T_j = T_a + P_d \theta_{JA} \quad \text{(Eq. 5-7)}$$

Sometimes the junction-to-case thermal resistance (θ_{JC}) is not given, and a maximum case temperature is specified. The above equations can be arranged to calculate the case-to-ambient thermal resistance (θ_{CA}) as follows:

$$\theta_{CA} = \frac{T_c - T_a}{P_d} \quad \text{(Eq. 5-8)}$$

Given these equations, let us look at a typical example and make some calculations. A pocket calculator may be used to verify the figures or perform similar calculations for different thyristor-heat-sink combinations.

Example:

An SCR installed on a heat sink is used to switch a dc load current of 10 A from a dc power source of 120 V. The maximum ambient temperature in the operating environment is 50° C, or about 122° F. Specifications for the SCR and heat sink are given as follows:

A. SCR
$I_{T(RMS)}$ for $T_c = 100°$ C: 16 A
V_{RROM}: 400 V
V_{DROM}: 400 V
P_d for $I_{T(DC)} = 10$ A: 13 W (From Graph)
θ_{JC}: 2° C/watt
T_c for $I_{T(DC)} = 10$ mA: 99° C (From Graph)
T_j: 125° C
B. Heat Sink
θ_{SA}: 2.05° C/watt
C. Assumed θ_{CS} using silicon thermal compound:
0.5° C/watt

Verify that this SCR can safely be operated in the given environment.

Solution:

1. In addition to the stated thermal ratings, the SCR manufacturer provided additional operating data in the form of graphs. As a first step, the power dissipation (P_d) of the SCR for a dc current of 10 A was found to be 13 W with the aid of the graph designated "On-State Power vs dc and ac RMS Current." The 13-watt value corresponds to a V_T of about 1.3 V, a reasonable value for this current level.

The maximum power dissipation for this SCR operating in an ambient temperature of 50° C and installed on the above heat sink is computed as follows:

$$P_d = \frac{T_{j(max)} - T_a}{\theta_{JA}} = \frac{T_{j(max)} - T_a}{\theta_{JC} + \theta_{CS} + \theta_{SA}}$$

$$= \frac{125 - 50}{2 + 0.5 + 2.05} = \frac{75}{4.55} = 16.48 \text{ watts}$$

This calculation confirms that the SCR installed on the given heat sink can safely be operated as specified with ambient temperatures up to 50° C. The actual power dissipation is 13 W, whereas the maximum allowable power dissipation is about 16 W.

2. From the manufacturer's graph designated "Maximum Allowable Case Temperature vs On-state Current," the maximum case temperature (T_c) of the SCR for a dc current of 10 A was found to be 99° C. Let us compute the actual case temperature using the given operating conditions:

$$T_c = P_d(\theta_{CA}) + T_a = P_d(\theta_{CS} + \theta_{SA}) + T_a$$

$$= 13(0.5 + 2.05) + 50 = 83.15°\text{ C}$$

Again, we have confirmed that the SCR is being operated within safe limits. The actual case temperature is about 83° C, well under the maximum specified case temperature.

Repetitive-Pulse Conduction—Thyristors operating in the repetitive-pulse mode can accommodate peak power dissipation levels that are higher than the maximum allowable steady-state dissipation. However, if the thyristor can possibly be latched in the steady-state condition, thermal calculations should be made on the assumption of steady-state operation.

Average power dissipation and thermal capacitance are the major factors in analyzing repetitive-pulse conduction. Thermal capacitance is defined as the ability of a material to store heat during warm-up and to release heat during cooling before the next pulse. Usually related to the junction-to-case path, this thermal capacitance can prevent the thyristor from cooling between pulses. Without proper design, a buildup of heat can occur and destroy the thyristor. For this reason, steady-state thermal calculations based on average power dissipation are not accurate in describing thyristor heat-flow conditions.

Thermal-resistance calculations for repetitive-pulse conduction are quite complex and beyond the scope of this book. Fortunately, in most applications, the manufacturers' operating graphs contained in specification sheets can be used for determining maximum case temperature for a given set of operating conditions. For example, virtually all specification sheets for phase-control SCRs include graphs designated "Maximum Case Temperature vs Average Forward Current" for varying conduction angles. Sine-wave, square-wave, and dc current conduction are usually included as variables on these graphs. For those who are used to measuring ac load current in rms values, average current can be converted to rms by the following equation:

$$I_{T(RMS)} = 1.111 \, I_{T(AVG)} \qquad \text{(Eq. 5-9)}$$

Another operating graph included in most power SCR and triac specifications is "Maximum Power Dissipation vs Average Forward Current." This graph, given for sine-wave, square-wave, and dc conduction, may be used to determine maximum power dissipation for a given conduction angle.

Nonrepetitive-Pulse Conduction—A given

thyristor can conduct a very high current for a brief period of time. This characteristic provides an instantaneous overload capability which allows time for an orderly shut-down during an unusual circuit malfunction. However, such overloads may cause thermal fatigue or minute permanent damage to the thyristor. Usually, a manufacturer will specify that a thyristor will withstand up to 100 individual nonrecurrent overloads during the anticipated operating life. Note that the junction temperature must return to normal after each overload condition.

The *peak forward surge current* (one cycle, 60 Hz) rating, or I_{TSM}, is one way the manufacturers use to describe single-pulse conduction. For example, an SCR rated for an on-state current (I_T) of 35 A_{rms} may safely conduct up to 300 A_{rms} when it is switched on for one cycle of 60-Hz power.

Maximum nonrepetitive current surges of less than one cycle duration can be computed with the aid of a graph that shows a transient thermal impedance curve. Sometimes called "Transient Thermal Impedance vs Time," this graph defines the junction-to-case thermal impedance for short periods of overload conduction ranging from about 0.0001 to 10 seconds. The graph is usually based on an existing operating case temperature of 100°C. Some manufacturers also include curves for repetitive-pulse operation at varying duty cycles on the same graph.

Commercial Heat Sinks

Heat-sink design involves a combination of electrical, mechanical, and thermal considerations. The ideal design provides for maximum cooling efficiency with minimum size. Commercial organizations have expended a considerable amount of time and money for research in developing optimum heat sinks for virtually all semiconductor devices, including thyristors. Figure 5-7 shows typical commercial heat sinks available for thyristor applications. A representative list of heat-sink manufacturers is given in Chart 5-1.

Heat-sink manufacturers furnish detailed technical data concerning their products. This generally

Chart 5-1. Representative Heat-Sink Manufacturers.

International Electronic Research Corp. (IERC)
135 West Magnolia Blvd.
Burbank, CA 91502

The Staver Co., Inc.
41-51 North Saxon Ave.
Bay Shore, NY 11706

Thermalloy, Inc.
P.O. Box 34829
2021 West Valley View Lane
Dallas, TX 75234

Wakefield Engineering, Inc.
60 Audubon Road
Wakefield, MA 01880

includes mechanical specifications and thermal-resistance ratings or graphs that show power dissipation vs temperature rise above ambient. Also, most manufacturers will provide, upon request, application notes concerning the use of their heat sinks.

Commercial heat sinks may be divided into three general categories relating to power-dissipation levels and type of thyristor:

1. Natural Air Convection and Radiation — Low- to Medium-Power Applications
2. Forced Air Cooling — Medium- to High-Power Applications
3. Forced Liquid Cooling — Medium- to High-Power Applications

No fine line can be drawn between these categories. The required power dissipation, operating environment, and individual thyristor characteristics dictate the proper type of heat sink. For example, an SCR being used to switch a load current of 15 A may require a heat sink cooled by forced air when it is operated in an environment with a high ambient temperature. The same SCR could be used with a heat sink cooled by natural air convection and radiation if the ambient temperature is maintained at a low level.

Heat sinks in each category possess individual

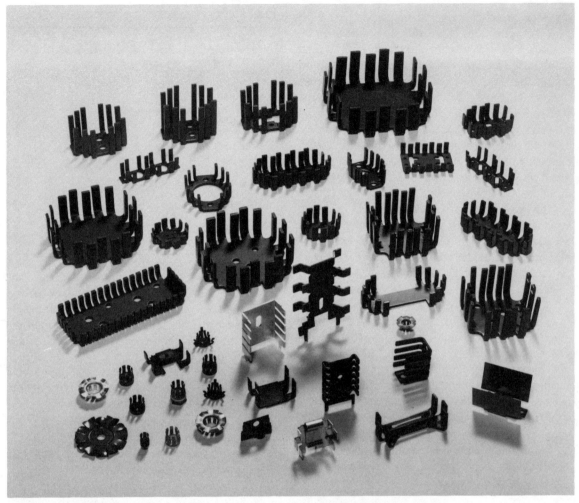

Fig. 5-7. Typical commercial heat sinks (courtesy of International Electronic Research Corp.).

electrical and mechanical characteristics. For example, electrical isolation between the thyristor and heat sink is provided by some models. A careful analysis of the specific heat-sink requirement will aid in selecting the proper heat sink.

RADIO FREQUENCY INTERFERENCE

Thyristor switching circuits often act as miniature radio transmitters, producing electromagnetic signals that interfere with radio receivers and other electronic equipment. These unwanted signals are called radio frequency interference (rfi). In some instances the reciprocal is true—externally generated electromagnetic energy can cause malfunctions and false triggering of thyristor control systems.

The Federal Communications Commission is the primary agency that governs and supervises the radiation of electromagnetic energy in the United States. This radiation includes both *intentional* signals from radio, television, and other communications equipment and *unintentional* man-made interfering signals. In general, any signal radiation that interferes with the operation of radio, television, or other electronic devices is expressly forbidden.

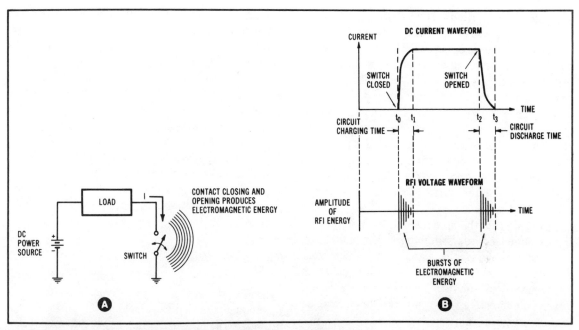

Fig. 5-8. Transient response of simple dc circuit. (A) Circuit. (B) Waveforms.

Adequate rfi suppression in thyristor control systems begins in the initial circuit design phase. Preplanning for proper shielding and filtering is usually adequate to reduce rfi levels to insignificant or acceptable levels. However, rfi-suppression techniques represent a broad mixture of practical experience and theoretical knowledge. The newcomer to thyristor circuit design should be prepared to seek assistance from experienced circuit designers or make use of available texts and application notes dealing with this subject. We will cover the more important aspects of rfi suppression in this section.

Basic Theory

The abrupt switching of current in either dc or ac circuits produces a transient response with accompanying pulses, or bursts, of electromagnetic energy. Sometimes this phenomenon is referred to as electrical noise.

Consider the simple dc switching circuit in Fig. 5-8, which could employ either a manually operated switch or the contacts of a relay. The closing or opening of these contacts causes a sharp rise or fall of current. During each transition, a short period of time is required for the current to reach a maximum level or fall to zero. This "charging" and "discharging" time is primarily a function of the reactive and resistive characteristics of the circuit. It is during these transition periods that high-frequency oscillations, and therefore electromagnetic energy, are produced. Theoretically, this energy consists of components with frequencies that extend to infinity and with amplitudes that decrease as the frequency increases. In practical circuits, signals with frequencies of hundreds of megahertz can be detected with sensitive radio receivers.

The fast turn-on and turn-off characteristics of thyristors are similar to those of the mechanical switch in Fig. 5-8. An SCR, for example, possesses a turn-on time on the order of a few microseconds. In a switching circuit, the SCR produces a transient response that gives rise to high-frequency oscillations up to several megahertz. If not properly suppressed, this will interfere with the reception of radio signals in this frequency range.

Thyristor ac switching circuits also exhibit rfi

characteristics, particularly when the ac voltage is switched during the positive or negative half-cycles. For example, a triac being switched on during each half-cycle with a firing angle of 90° will generate brief bursts of rfi twice each cycle. With a 60-Hz power source, the pulsed rfi occurs at a fundamental frequency of 120 pulses per second. This action is illustrated in Fig. 5-9. These repetitive bursts of rfi signals occur at many individual frequencies, producing interfering signals up to several megahertz. For example, a triac lamp-dimmer circuit may cause interference to am broadcast-band reception.

In addition to the fundamental frequency of rfi bursts or pulses, the switching circuit generates higher-order harmonic signals of appreciable amplitude. The triac circuit in Fig. 5-9 generates harmonic signals in the audio-frequency range which may interfere with telegraph, telephone, or hi-fi audio systems.

Conducted and Radiated Rfi

Electromagnetic energy generated by a thyristor switching circuit may be transmitted to the outside world by two methods: free-space radiation of electromagnetic waves, or conduction via power or signal/control lines. Radiation occurs when portions of the thyristor control circuit act as antennas. For example, long unshielded wires connecting the thyristor to the load may serve as efficient antennas for the lower-frequency electromagnetic energy.

Power and signal/control lines may also provide excellent conduction paths for rfi or electrical noise. When these lines are connected to other equipment, the interfering signals can produce serious malfunctions. In other instances, thyristor power or control lines installed in common cable ducts may permit electrical noise to be induced into other cables.

Rfi Suppression

The suppression of rfi in thyristor control systems is based on two individual but related techniques, filtering and shielding. Low-pass filters attenuate the higher-frequency components of the rfi energy, thus reducing both radiated and conducted rfi. Properly designed metallic equipment enclosures or cabinets act as shields and prevent radiation from leaving or entering the thyristor circuits. Common to both filtering and shielding is a well designed grounding system or common ground

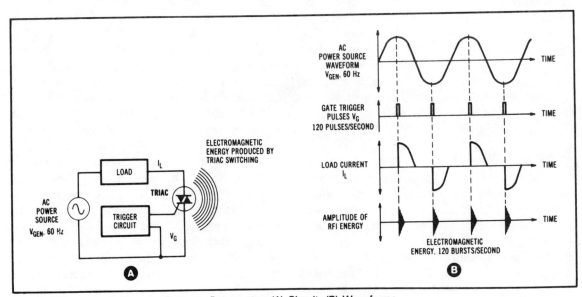

Fig. 5-9. Triac ac switching circuit as an rfi generator. (A) Circuit. (B) Waveforms.

bus. This is extremely important in reducing radiated rfi energy.

Most radiated and conducted rfi energy can be suppressed at the source by incorporating a low-pass LC filter in the load-switching circuit of the thyristor. Filter design for thyristor circuits is normally based on eliminating interfering signals with frequencies up to several megahertz. Above these frequencies, the rfi levels are usually too low to cause interference to radios or other electronic equipment. Many thyristor low-pass filters are designed for a corner, or cutoff, frequency of about 50 kHz. This permits the use of small, inexpensive filter components. Figure 5-10A illustrates this technique.

The equivalent rfi model in Fig. 5-10B shows the triac as a high-frequency voltage source with many harmonics. Without the LC filter, this voltage source would be connected directly to the load and associated power lines. Thus, rfi energy could be radiated from this circuit as well as being conducted to other electronic equipment via common power lines. However, the LC filter isolates the high-frequency voltage source from the load and power lines. The inductor presents a high inductive reactance, while the capacitor acts as a low capacitive reactance. Each component attenuates the rfi level by approximately 20 dB per decade below the level at the cutoff frequency. The combined LC filter circuit produces a rolloff attenuation of about 40 dB per decade. Since the rfi signal decreases in amplitude at a rate of about 20 dB per decade, the total attenuation with the LC filter is about 60 dB per decade. This is illustrated in Fig. 5-10C—the rfi levels in this example are adequately suppressed in the am broadcast band of 535 kHz to 1.605 MHz.

The cutoff, or corner, frequency (f_c) of a simple low-pass filter is related to the load resistance and the L or C reactive components as follows:

$$R_L = X_L = X_C \quad \textbf{(Eq. 5-10)}$$

Since

$$X_L = 2\pi f_c L \quad \textbf{(Eq. 5-11)}$$

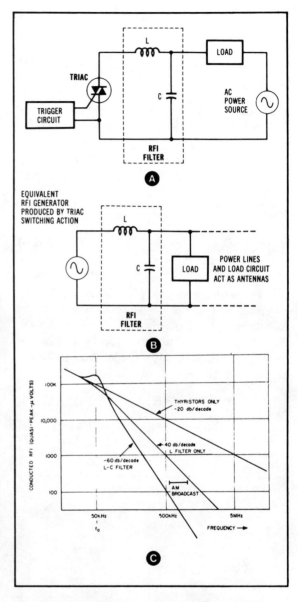

Fig. 5-10. Typical thyristor rfi filter circuits (courtesy of General Electric Co.). (A) Triac switching circuit with LC low-pass filter. (B) Equivalent high-frequency circuit. (C) Theoretical thyristor-circuit noise spectrum.

and

$$X_C = \frac{1}{2\pi f_c C} \quad \textbf{(Eq. 5-12)}$$

then

$$R_L = 2\pi f_c L = \frac{1}{2\pi f_c C} \quad \text{(Eq. 5-13)}$$

From these equations, we can develop the key equations for determining the values of L and C for a simple LC low-pass filter:

$$L = \frac{R_L}{2\pi f_c} \quad \text{(Eq. 5-14)}$$

and

$$C = \frac{1}{2\pi f_c R_L} \quad \text{(Eq. 5-15)}$$

If a corner frequency of 50 kHz is selected for our filter design, the values of L and C can be calculated by the following simplified equations. Note that 50 kHz represents approximately one decade of frequency below the low end of the am broadcast band.

$$L = 3.183\, R_L \quad \text{(Eq. 5-16)}$$

$$C = \frac{3.183}{R_L} \quad \text{(Eq. 5-17)}$$

where,
R_L is in ohms,
L is in microhenries,
C is in microfarads.

The corner frequency (f_c) also represents the resonant frequency of the equivalent LC series circuit. When combined with the low resistance values of the thyristor circuit, the resulting low Q produces a slight buildup of voltage across the capacitor at the corner frequency. This will be remembered as the *Q rise of voltage* in a series LC resonant circuit. It is shown as a slight hump at 50 kHz in the curve in Fig. 5-10C.

The simple LC low-pass filter is usually adequate for most thyristor switching circuits involving low to medium power levels and resistive loads. The triac lamp dimmer in Fig. 5-11 is a typical circuit of this category. Higher-power thyristor switching with reactive loads may require more complex rfi filters or electromagnetic shielding, or a combination of the two. Figure 5-12 illustrates how shielding and related bypass capacitors help in containing rfi energy within the thyristor switching circuit.

Rfi Interaction with Thyristor Circuits

The presence of externally generated elec-

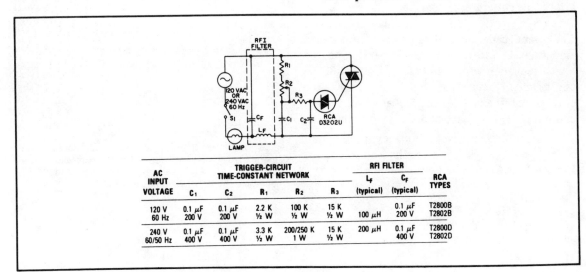

AC INPUT VOLTAGE	TRIGGER-CIRCUIT TIME-CONSTANT NETWORK					RFI FILTER		RCA TYPES
	C_1	C_2	R_1	R_2	R_3	L_F (typical)	C_F (typical)	
120 V 60 Hz	0.1 µF 200 V	0.1 µF 200 V	2.2 K ½ W	100 K ½ W	15 K ½ W	100 µH	0.1 µF 200 V	T2800B T2802B
240 V 60/50 Hz	0.1 µF 400 V	0.1 µF 400 V	3.3 K ½ W	200/250 K 1 W	15 K ½ W	200 µH	0.1 µF 400 V	T2800D T2802D

Fig. 5-11. Triac light dimmer circuit with rfi filter (courtesy of RCA).

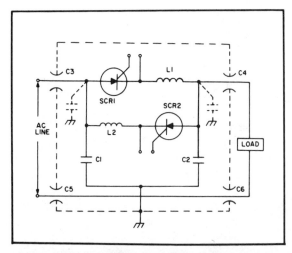

Fig. 5-12. Rfi filtering and shielding (courtesy of General Electric Co.).

tromagnetic radiation or electrical transients may interfere with the proper operation of a thyristor switching circuit. Unshielded control wires in the thyristor trigger circuit or long unprotected power lines connected to the thyristor anode or main terminal connector may act as antennas. This antenna effect may permit external rfi to trigger the thyristor into conduction at an undesired time.

One obvious solution to this problem is the use of shielded or twisted cables for the control lines in the trigger circuit and low-pass filter in the main current-switching circuit of the thyristor. In general, all the precautions taken to suppress internally generated rfi are also applicable to external rfi interaction. In addition, the use of low-impedance control lines and optoisolator trigger circuits will aid in eliminating problems with false triggering.

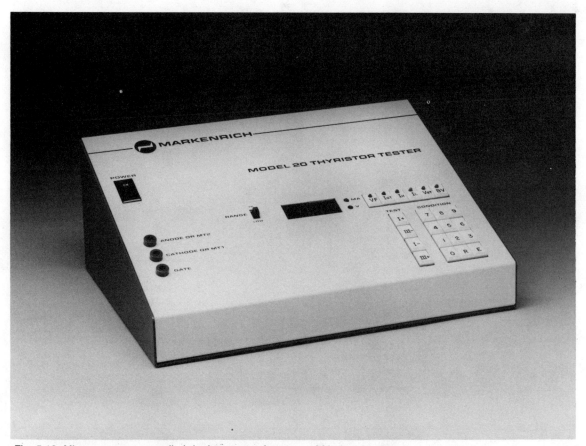

Fig. 5-13. Microprocessor controlled thyristor tester (courtesy of Markenrich Corp.).

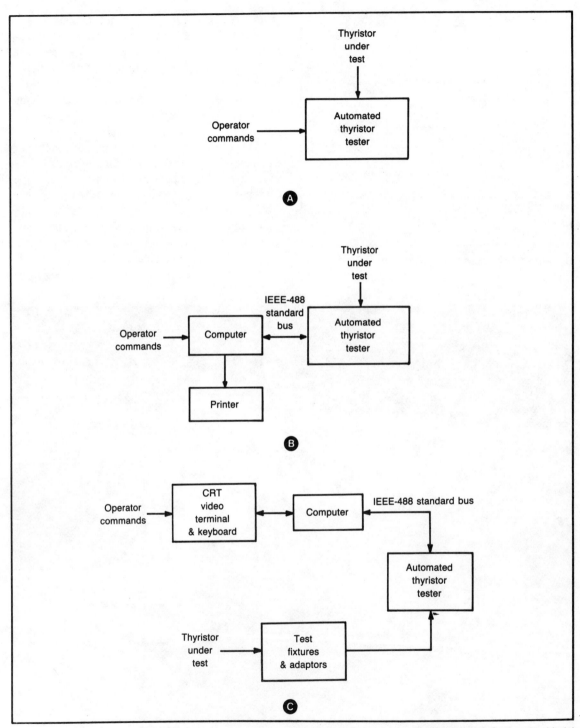

Fig. 5-14. Test configurations for using microprocessor controlled thyristor testers such as the Markenrich Corporation Model 20 Thyristor Tester. (A) Stand alone. (B) Computer assisted. (C) Automated Test Equipment (ATE).

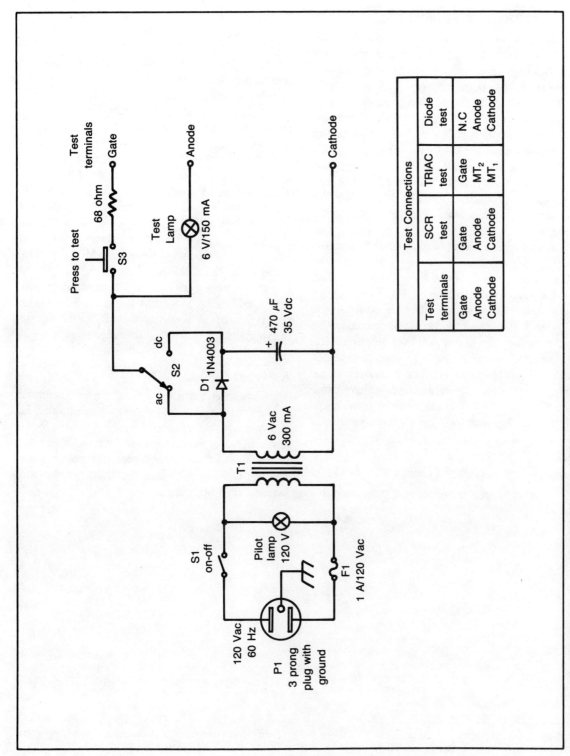

Fig. 5-15. Simple thyristor tester for "GO-NO-GO" tests of SCRs, triacs and diodes.

TESTING THYRISTOR DEVICES

Our final topic for planning the power control system is that of testing thyristors. Some means must be provided for installation and maintenance personnel to test thyristors. Much of this testing can be handled with the use of a good multimeter and oscilloscope. In most instances, the thyristor is either good or defective—multimeter continuity and voltage checks can isolate many problems while oscilloscope analysis of gate trigger signals will help to show why the thyristor is not operating properly. Detailed test procedures must be documented for the control system in order to ensure fast, prompt maintenance.

In some instances, you will need to test the thyristor to determine if it meets manufacturer's specifications. For example, some thyristors—such as SCRs and triacs, will self-trigger or oscillate under certain operating conditions. Reverse blocking voltage tests are important in determining this condition. It is almost impossible to determine this type of problem with simple multimeter checks. Also, to simulate this condition in an operational system may be an expensive and prohibitive situation. Therefore, some means must be provided to remove the thyristor from the control system and then run individual tests.

Commercial Thyristor Testers

Commercial testers are available for checking a wide range of thyristors such as SCRs, triacs, rectifiers, diacs and similar devices. Figure 5-13 shows a thyristor tester which operates much like a digital voltmeter. It can be a stand-alone unit or integrated into an automated test unit. This tester covers dc parametric tests for gate trigger voltage and current, holding current, latching current, forward and reverse blocking voltage, and forward voltage drop. Thyristors with gate currents to 400 mA, and blocking voltages to 2000 V can be tested for proper triggering action in up to four quadrants of operation. These testing techniques are useful for determining if a device will self-trigger or oscillate under certain conditions.

The tester is built around a microprocessor and 2K ROM (or 2000 bits of Read-Only-Memory). The tester, with its IEEE-488 interface, can be integrated with a computer for automated or programmed testing procedures. Figure 5-14 shows the types of test configurations available for microprocessor controlled thyristor testers.

A Simple "GO-NO-GO"

Sometimes a simple thyristor tester is required to check for "GO-NO-GO" operation. Figure 5-15 shows a circuit for a tester which provides for ac or dc checks of SCRs, triacs and diodes. However, proper use of this simple tester requires that you become familiar with specifications for the particular thyristor being tested.

Chapter 6

Automation, Robots, and Thyristors

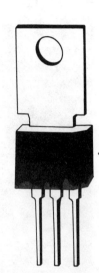

THE DEVELOPMENT OF DIGITAL ELECTRONICS and integrated-circuit technology over the past twenty years has created a new industrial revolution. Automated production machines, controlled by digital logic circuits, are being used on assembly lines to assist in the manufacture of automobiles, tv sets, and other mass-produced products. The petroleum and chemical industries employ automated control systems in refining and chemical-processing plants. These sophisticated *automated machines*, sometimes referred to as *robots*, are being used in virtually every phase of industry. Figure 6-1 illustrates automated machines used in a production environment.

In many automated control systems, the thyristor is the final link in performing the work function. Thyristors turn on motors, valves, welding machines, and other equipment. In this chapter, we will examine some of the more important aspects of this new technology and the interfaces required for thyristor firing circuits.

AUTOMATED PRODUCTION AND PROCESSING

The production and distribution of energy, materials, and finished products rely more and more on automated control systems. These efficient robots are used in mass-production line operations that involve drilling, milling, processing, final assembly, packaging, and printing. Food processing, automobile assembly, consumer electronics, and other types of factories are employing a wide variety of automation techniques to produce quality products at competitive prices. Even drafting and electronic design functions are being automated by some major companies.

A Historical Overview

No discussion of automated and digital systems would be complete without some historical background. Table 6-1 provides a short review of some of the notable achievements in this field.

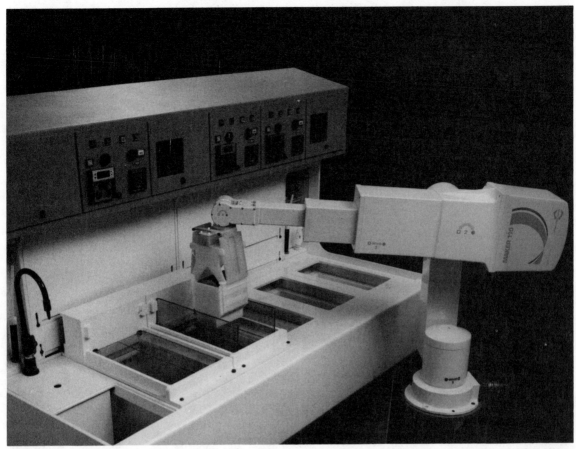

Fig. 6-1. Automated "robot" machine being used in a semiconductor wafer etching application (courtesy of United States Robots).

The introduction of the *digital computer* in the 1940s paved the way for automated information- and data-processing systems. These early computers provided an efficient means for making accounting and scientific computations such as inventory control, finance and personnel records, model studies, and statistical analyses. These *first-generation* computers were giants that required large installation areas, air-conditioned environments, and considerable electrical power. Employing thousands of vacuum tubes and costing a million dollars or more, these fragile machines were too expensive and unreliable for automated production-line applications.

A major milestone was reached in 1947 with the invention of the transistor. Requiring only a fraction of the size and power demanded by the vacuum tube, the inexpensive and reliable transistor was quickly incorporated into digital control and computing systems. The *second-generation digital computer,* a small solid-state system, began appearing in production lines and shop areas to perform limited automated functions.

The development of integrated circuits in the early 1960s further reduced the size and cost of digital systems. Some of the early IC "chips" contained up to 12 individual logic gates, or the equivalent of about 50 transistors, diodes, and resistors. Sometimes referred to as *small-scale integration* (SSI), these tiny devices began to appear in many automated digital control systems. Digital computers fabricated with the ICs were known as

Table 6-1. Some Historical Events Concerning the Development of Automation and Computing Systems.

500 B.C.	Abacus, invented by the Chinese. Still in use in some parts of the world.
1621 AD	Sliderule, Oughtred in England.
1643	Pascal's Arithmetic Machine, France. Used a series of wheels with numbers 0-9 on each wheel for arithmetic operations.
1725	Punched paper tape for controlling cloth weaving machines, Bouchon in France.
1834	Babbage's Analytical Engine, England. Used internally stored program contained in punched cards. This device is considered to be the forerunner of modern computers.
1885	Burroughs Adding Machine, USA.
1890	Hollerith Punched Card Machines, USA. Led to punched-card accounting machines and calculators. Forerunner of IBM Corporation.
1904	First vacuum tube, Fleming in England.
1917	First desk calculator, USA.
1940	First electrical analog computers, Parkinson and Lovell, Bell Telephone Laboratories, USA.
Early 1940s	First electrical digital computer, Aiken's Mark 1, USA. Used over 3000 relays and weighed 5 tons.
1946	ENIAC (Electronic Numerical Integrator And Calculator), Echert and Mauchly, USA. First electronic digital computer. Used 18,000 vacuum tubes. Did not include an internally stored program.
	Von Neumann Internally Stored Program Concept for electronic digital computers.
1947	First transistor, Bell Telephone Laboratories scientists, USA.
1949	EDSAC, England. First electronic digital computer to use internally stored program concept.
Late 1940s and Early 1950s	First commercial electronic digital computers, UNIVAC 1, IBM 701, and IBM 704, USA. Used vacuum tubes and magnetic core memories. Considered to be first generation computers.
Late 1950s	First commercial transistorized electronic digital computers. Considered to be second generation computers.
Early 1960s	First integrated circuits, Texas Instruments, USA. Led to third generation computers and numerically controlled machines.
Late 1960s	Large-scale integration (LSI). Up to 1000 logic gates per chip.
Early 1970s	Very-large-scale integration (VLSI). Up to 50,000 logic gates per chip.
1971	First microprocessor or "computer-on-a-chip," Intel, USA. This development along with VLSI integrated circuits led to fourth generation computers, pocket calculators, home computers, and other modern devices.

third-generation types.

As design and manufacturing techniques improved, the capability and number of logic gates per chip were substantially increased. By the late 1960s, large-scale integration (LSI) techniques provided for up to 1000 logic gates per chip. This made possible the familiar and inexpensive, but powerful, pocket calculators as well as other small digital devices.

The 1970s ushered in two major advances in electronics technology, the microprocessor and very-large-scale integration (VLSI). The number of individual logic gates per chip was increased to over 50,000. Using VLSI techniques, the Intel Corp. developed the first microprocessor, or "computer-on-a-chip," in 1971. By the mid-1970s many companies were producing microprocessors with varying capabilities. Figure 6-2 shows a typical microprocessor. In addition, VLSI technology permitted the fabrication of high-capacity solid-state memories, special-purpose digital circuits, and combined analog-digital integrated circuits. (The *analog* computer processes continuous quantities, such as changes in voltage or resistance, whereas the *digital* computer processes discrete numbers.) Many computer specialists refer to the microprocessor-based computer with VLSI memories as the *fourth-generation* stage of development. Microprocessors are used to control the actions of robot machines. The digital ac servo robot shown in Fig. 6-3 employs microprocessors, drive assemblies, and pneumatic mechanisms for smooth movement and precise applications.

Numerically Controlled Machines

The numerically controlled (N/C) machine is defined as a machine tool controlled by a sequence of numerical instructions. Automated drilling or milling machines capable of operating from a preprogrammed set of instructions fall within this category.

The original concept for the N/C machine dates back to the late eighteenth century during the Industrial Revolution. Crude N/C devices that used punched paper tapes containing coded instructions were employed with steam-driven cloth-weaving machines. Widespread use of N/C machines, however, were not initiated until the 1960s. The invention of the transistor and the development of digital

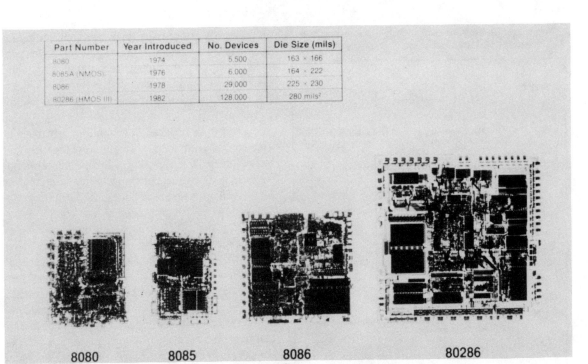

Fig. 6-2. A typical microprocessor or "computer-on-a-chip" (courtesy of Intel Corp.).

Fig. 6-3. A digital ac servo robot employing microprocessors for precise control applications (courtesy International Robomation/Intelligence).

electronics techniques made possible the efficient and reliable N/C machine.

Figure 6-4 provides a basic block diagram of an N/C drilling machine. Used in steel-fabrication plants, this type of N/C machine automatically drills holes at preprogrammed locations in steel beams. Industrial and commercial building construction uses many beams of equal or similar sizes and fast, efficient drilling of the required holes can be accomplished with N/C machines.

Actual N/C machine configurations vary from one model to the next. Some N/C machines utilize a standard eight-hole paper tape for inputting the preprogrammed work functions to the computer. Other N/C machines employ magnetic tape in cassettes for program input operations. In some instances, the paper or magnetic tape is prepared by a computer when the customer's order form is received at the plant. Other, more modern N/C machines incorporate a microprocessor and a general utility stored program. In these cases, the N/C operator manually inputs data via a keyboard in order to program and activate the N/C machine.

Some N/C machines operate on a point-to-point positioning basis—the tool or the work piece is positioned at discrete points for the required work function. The N/C drilling machine in Fig. 6-4 illustrates this concept. Since this approach represents a minimum of investment and manpower, most of the N/C machines are of this type.

Other, more complex N/C machines perform continuous-path work functions. This type of operation may involve a three-dimensional work pattern, such as contouring a metal surface or making continuously welded seams. These N/C machines usually employ microprocessor-based digital control systems with extensive internally stored programs. Many factories use these complex robots in final assembly operations.

As stated earlier, thyristors are frequently the last control link in automated production machines. The N/C drilling machine in Fig. 6-4 would use thyristors to control the drill motors and the horizontal-drive motor. In a typical configuration,

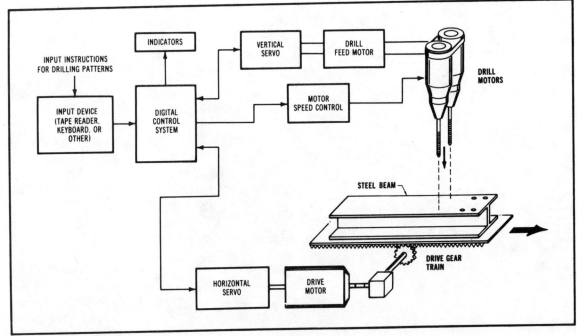

Fig. 6-4. Block diagram of numerically controlled drilling machine.

high-current triacs would be most appropriate for this task. After reviewing some microprocessor fundamentals, we will investigate the interface between digital electronics and thyristor firing circuits.

MICROPROCESSORS AND MICROCOMPUTERS

A microprocessor is a large group of related digital circuits, all contained within one IC chip. The tiny chip, typically 1/4 inch square and 1/100 inch thick, is usually encapsulated in a plastic dual in-line package (DIP) with 40 or more connector pins. When the microprocessor is interconnected with input and output devices, external memory, and a power supply, the resulting configuration is called a microcomputer. The external memory and input/output control circuits may be contained within other high-density chips.

Figure 6-5A shows a simple block diagram of a microcomputer system along with a typical microcomputer board (Fig. 6-5B) containing a microprocessor, input/output chips, random access memory (RAM), and read only memory (ROM). Equipment such as this is employed in N/C machines or other digital electronic systems. With a stored program in the external memory, the basic actions of this microcomputer are as listed in the following sequence. The microprocessor is at the center of the action, "running the show."

1. External signals (data or command) are generated and presented to the input section.

2. The input section notifies the microprocessor that input signals are available.

3. The microprocessor requests the input section to accept the input signals and transfer them to the microprocessor data input terminals.

4. The microprocessor examines the input signals to determine the proper course of action. If further logical or processing operations are required, the microprocessor checks with the stored program in the memory section for the required course of action. In some instances, the microprocessor will simply transfer the input data to the memory section for storage and later processing.

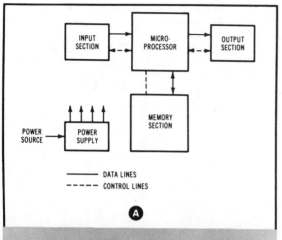

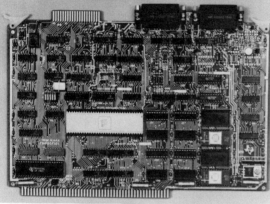

Fig. 6-5. Configuration of a microcomputer system (courtesy of Texas Instruments Inc.). (A) Simplified block diagram. (B) Typical microcomputer module.

5. When the microprocessor completes the required logical or processing operations, the output section is notified that output data is available for a specified output device.

6. The output section, in turn, requests the specified output device to acknowledge when it is ready to accept the output data.

7. When a ready signal is generated by the output device, the microprocessor transfers the output data to the output section for subsequent delivery to the output device.

8. While the output data is being transferred to the output device, the microprocessor checks with the stored program in the memory for further instructions. If no further action is required at this time, the microprocessor remains idle until new input data is presented to the input section.

Binary Numbers

All microprocessors operate with a binary number system involving two digits—0 (zero) and 1 (one). The decimal number system is too inefficient and unwieldy for use in microprocessors and other digital electronic systems. Signals of 0 V and +5 V (approximately) are used in most digital microprocessors for the binary 0 and 1 digits. This results in a simplified design, since each transistor in the logic gates and other digital circuits is operated either in the cutoff or saturated condition.

In order to handle the large numbers associated with microprocessor operations, blocks of *bits* (binary digits) are employed in a positional-notation scheme. This results in a binary (base 2) numbering system. Since we are accustomed to expressing decimal (base 10) numbers in this matter, it is easy to visualize binary numbers in a similar fashion. For example, the position of a digit in the decimal numbering system indicates that this number will be multiplied by 10 to a particular power. Let us examine the number 876. This is equivalent to:

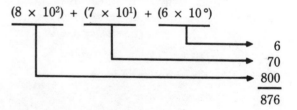

$$(8 \times 10^2) + (7 \times 10^1) + (6 \times 10^0)$$

$$\begin{array}{r} 6 \\ 70 \\ 800 \\ \hline 876 \end{array}$$

Note that 10 to the zero power is defined as 1. In fact, any number raised to the zero power is 1.

In a similar manner, any quantity can be expressed in a binary numbering system by using the following format:

$$\ldots (N_4 \times 2^4) + (N_3 \times 2^3) + (N_2 \times 2^2) + (N_1 \times 2^1) + (N_0 \times 2^0)$$

In this expression, each of the N factors (N_0, N_1, etc.) can be either 1 or 0.

For example, the number 29 in the decimal numbering system can be expressed in the binary number 11101. This is illustrated in Fig. 6-6 in terms of positional notation.

Nibbles, Bytes, and Words

The first microprocessor, the Intel 4004, is based on a standard 4-bit word. Table 6-2 shows the range of numbers possible with four bits. These combinations of bits from 0000 to 1111 are also compared to the *decimal* and *hexadecimal* (base 16) equivalent numbers. We will discuss hexadecimal numbers later.

The 4-bit word permits us to express 16 unique values (from 0 to 15). The Intel 4004 can accept one 4-bit word for each input operation. Since most numeric data involves numbers larger than 15, we can use additional 4-bit words can be used to express 256 different numbers (from 0 to 255). The largest number (255) is equivalent to 1111 1111. This is illustrated in Fig. 6-7.

Further advances in microprocessor technology led to the development of 8-bit, 16-bit, and 32-bit "machines." This concept made faster and more efficient input/output and numerical processing operations possible. Thus, large numbers can be handled more easily with the larger microprocessors. In microprocessor terminology, the 4-bit word is referred to as a *nibble*, and the 8-bit word is called a *byte*.

Table 6-2. A 4-Bit Binary Code with the Equivalent Hexadecimal and Decimal Codes.

4 Binary Bits	Equivalent Hexadecimal Code	Equivalent Decimal Code
0000	0	0
0001	1	1
0010	2	2
0011	3	3
0100	4	4
0101	5	5
0110	6	6
0111	7	7
1000	8	8
1001	9	9
1010	A	10
1011	B	11
1100	C	12
1101	D	13
1110	E	14
1111	F	15

Hexadecimal Numbers

Microcomputer programmers and operators generally express microprocessor programs in terms of the hexadecimal numbering system. This is convenient because each 4-bit word permits counting from 0 to F (refer to Table 6-2). By using two 4-bit words, it is possible to count from 00 to FF, where FF is the equivalent of 255 (base 10). Similarly, using three 4-bit words extends the range of numbers to FFF (base 16), or 4095 (base 10). We can verify this by using the following:

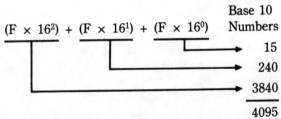

In the above example, note that F (base 16) is equal to 15 (base 10).

The hexadecimal number FFF can be expressed in binary notation. In fact it is easy to convert from hexadecimal to binary notation by merely expressing each hexadecimal digit in binary form. For example, the following hexadecimal numbers

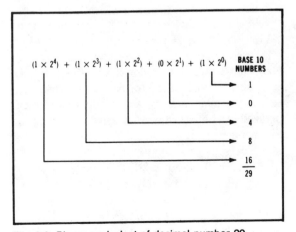

Fig. 6-6. Binary equivalent of decimal number 29.

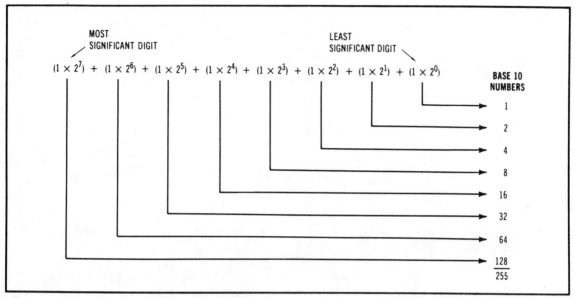

Fig. 6-7. Binary equivalent of decimal number 255.

converted to binary show the ease of this conversion:

```
FFF  ───────▶  1111 1111 1111
9BE  ───────▶  1001 1011 1110
28F  ───────▶  0010 1000 1111
```

Conversion from binary to hexadecimal form is a similar and easy process.

Binary and hexadecimal numbers can be used in arithmetic operations such as addition, subtraction, multiplication, and division. In fact, binary arithmetic circuits in microprocessors are relatively simple. In contrast, decimal arithmetic circuits for computers would be extremely difficult to implement. Unfortunately, space limitations will not permit a full review of binary arithmetic operations. For those who are interested in delving further into the techniques for handling binary and hexadecimal numbers, excellent texts are available at book stores and electronics parts houses.

Microprocessor Organization and Functions

Most of the circuits within the microprocessor chip are digital logic gates that perform logical operations on data received from the input devices. These logic gates also decode instructions received from the external memory and carry out the required actions contained in the instructions. Figure 6-8 shows a block diagram of a typical 8-bit microprocessor connected to external memory and an input/output device. A power supply (not shown) is the only item required to complete the basic microcomputer system.

Prior to using this microcomputer system for its intended application, we must input a program (a precise set of operating instructions) into the external memory. This program, which reflects the required processing or work functions in minute detail, is usually prepared in binary coding called *machine language*. In some instances, the program is permanently placed in a memory device referred to as Read Only Memory (ROM). This technique is used for mass-produced microcomputer systems in which the program is never changed. Input data or changing information is usually stored in temporary memory devices called Random Access Memory (RAM). A typical microcomputer memory unit may contain both ROM and RAM devices. VLSI construction techniques are used to fabricate solid-state memory chips for both ROM and RAM.

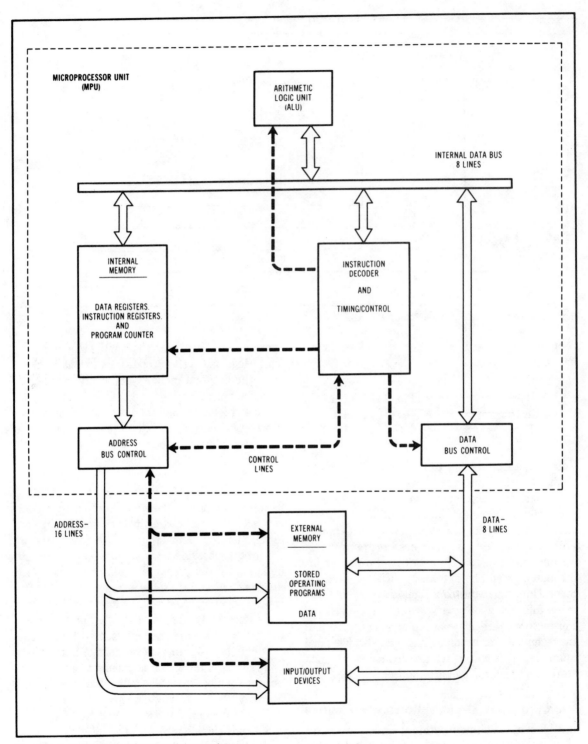

Fig. 6-8. A typical 8-bit microprocessor system.

Chart 6-1. Basic Categories of Instructions Used in Microprocessors.

Data Movement or Transfer
MOVE (Internal Transfers)
STORE (MPU to Memory)
LOAD (Memory to MPU)
INPUT (Input Device to MPU)
OUTPUT (MPU to Output Device)

Arithmetic and Logic Operations
ADD
SUBTRACT
MULTIPLY
DIVIDE
ABSOLUTE VALUE
NEGATE

LOGICAL AND
LOGICAL OR
LOGICAL NOT
LOGICAL EXCLUSIVE OR

Branch or Jump
UINCONDITIONAL JUMP
CONDITIONAL JUMP
JUMP TO SUBROUTINE

Compare and Test
Test for Zero Result
Test for Positive Result
Test for Negative Result

Register Shift Operations
SHIFT LEFT (One Bit)
SHIFT RIGHT (One Bit)

Modify Registers or Memory
INCREMENT (By One Bit)
DECREMENT (By One Bit)

At present, VLSI memory chips are capable of storing up to 256,000 bits or more per chip.

The typical microprocessor can perform up to 200 or more individual types of logical operations or instructions. Each instruction possesses a unique binary coding of 0s and 1s. Some instructions move data between the microprocessor and input/output devices or external memory. Other instructions perform arithmetic operations such as addition or subtraction. Chart 6-1 provides a list of the basic types of instructions used in microprocessors. Many variations of these basic instructions are available in the microprocessors available on the present market. Most programs can be prepared using the basic instructions listed. However, longer programs can be made more efficient and faster by using the special variations of these basic instructions which are offered by the various microprocessor manufacturers.

Input/output devices vary depending on the particular application. A combined keyboard and printer unit may be employed as an input/output device when human communication is required. Other input/output devices may include magnetic tape recorders (for long-term storage of mass data) or direct connections to communications circuits for computer-to-computer operation.

The basic operation of the simple microcomputer system in Fig. 6-8 is applicable to programmable pocket calculators, personal "hobby" computers, commercial or scientific computers, or modern-day N/C machines. For example, the N/C machine illustrated in Fig. 6-4 may incorporate a microcomputer as the digital controller.

One final note on microcomputers: most of these systems are more complex than the simple configuration shown in Fig. 6-8. For example, when more than one input/output device is employed, additional support digital ICs are required to interface these input/output lines with the microprocessor. Each microprocessor manufacturer markets a line of support ICs to allow flexibility in building the required microcomputer configuration.

THE THYRISTOR CONNECTION

Microprocessor-based or other digital logic systems can be used to control the firing of thyristors by a variety of methods. In many instances, the digital signal used to control the thyristor gate-trigger circuit is made available from a logic circuit such as an inverter stage or a NAND gate. Figure 6-9 shows a simplified approach to the thyristor control process.

An input command causes the microprocessor to initiate the control process. A keyboard, a manually operated switch, or an electrical signal

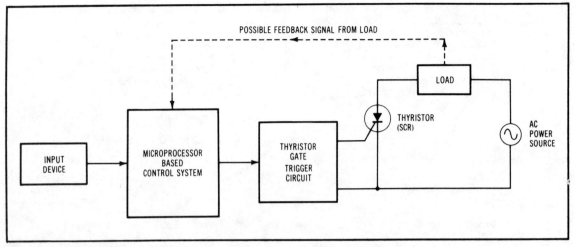

Fig. 6-9. Block diagram of a microprocessor-controlled thyristor switching system.

can be used for providing the command to start operation. At a designated point in the operating program, the microprocessor generates a firing signal that is transmitted to the thyristor gate-trigger circuit. The thyristor is switched on for the intended work function. When the work function is completed, the microprocessor must turn off the thyristor at the proper time. For example, a welding or milling operation must be terminated at the proper time. This action can be accomplished either by internal timing within the microprocessor control program or by a *closed-loop feedback* process. The closed-loop feedback concept requires that the load generate some form of electrical signal when the work function is completed. This signal is fed back to the microprocessor for switching off the thyristor.

The interface between the thyristor gate circuit and the microprocessor-based control system depends primarily on the following factors:

• Output control-signal characteristics, including amplitude, waveform, and internal impedance.
• Requirement for either zero-voltage firing or phase-delay firing angle.
• Electrical isolation required between thyristor and microprocessor-based control system.
• Required thyristor gate-trigger signal.

Digital Logic Gates

Logic gates are digital circuits that perform logical operations based on a binary (two-state) set

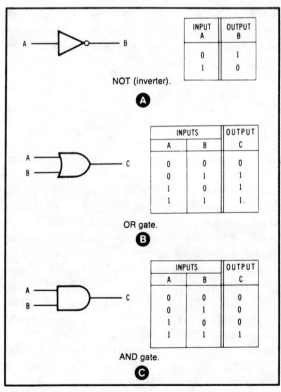

Fig. 6-10. The three basic logical operations.

170

of values. The two states are commonly referred to either as ones and zeros *or* as true and false statements. In terms of circuit operation, a logical one may be defined as a +5 V level, and a logical zero may be a 0 V level.

The three fundamental types of logical operations are the NOT, AND, and OR functions. Figure 6-10 illustrates the symbols and *truth tables* associated with each function. The truth tables show the output state for each combination of input states. For example, a 1 (+5 V) input to a NOT (inverter) stage produces a 0 (0 V) at the output and vice versa.

The NOT, AND, and OR digital logic circuits can be combined to form other basic logical operations such as the NAND and NOR functions. The NAND and NOR logic gates are simply AND and OR gates, respectively, whose output signals are fed into NOT stages. You may want to work out the logic diagrams and truth tables for these types of gates. Groups of logic gates and NOT circuits are also interconnected to form more complex digital elements such as *bistable multivibrators (flip-flops)*. The flip-flop acts as a one-bit memory device and is used in data registers, counters, solid-state memories, and other similar applications. Symbols for the NAND, NOR, and flip-flop are given in Fig. 6-11.

Digital logic circuits are based primarily on the transistor switch. However, some special digital logic devices employ various types of thyristors as the switching elements. In most instances, the transistor switch is either saturated or cut off. This mode of operation permits simple and highly reliable design of digital circuits.

Digital Logic Families

During the past twenty years, solid-state digital logic circuits have evolved from discrete-component configurations to integrated circuits utilizing either bipolar or field-effect transistor technology. The more common IC digital logic families are listed in Table 6-3. The CMOS and MOSFET digital ICs employ field-effect technology. The remaining logic families are based on bipolar technology. All of the logic families except ECL operate with saturated

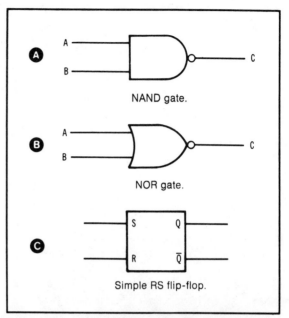

Fig. 6-11. The NAND, NOR, and flip-flop logic elements.

transistor switches in the conduction mode. Each logic family possesses inherent advantages and limitations.

Transistor-transistor logic (TTL) is the most popular logic family in use today. Sometimes referred to as the 74XX or 7400 series, this logic family features a wide variety of MSI and LSI ICs. These devices provide high-speed switching characteristics, reliable operation, and low cost. Inverters, NAND gates, flip-flops, multiplexers, decoders, and shift registers are typical of the 74XX series of ICs.

Figure 6-12 shows the two basic circuit configurations for inverters. Figure 6-12A shows one of the inverters in a 7404 hex inverter, and Fig.

Table 6-3. Integrated-Circuit Digital Logic Families.

CMOS	Complementary metal oxide semiconductor
DTL	Diode-transistor logic
ECL	Emitter coupled logic
I²L	Integrated injection logic
MOSFET	Metal oxide semiconductor field effect transistor
RTL	Resistor-transistor logic
TTL	Transistor-transistor logic

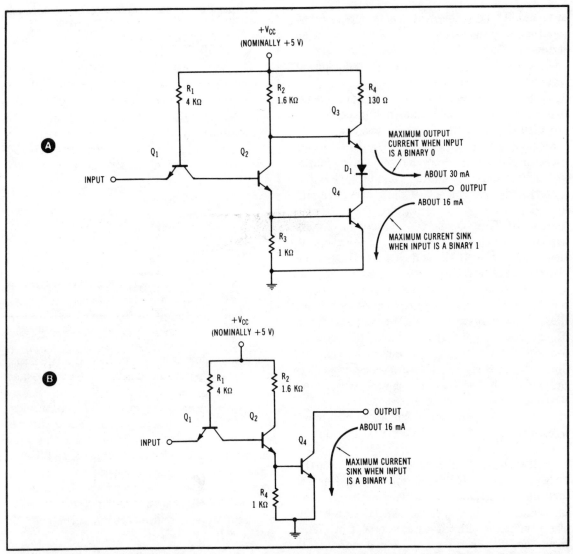

Fig. 6-12. Typical TTL inverters, or NOT, configurations. (A) Invertor circuit. (B) Invertor circuit with open-collector output.

6-12B shows one of the inverters with open-collector outputs in a 7405 hex inverter. (Actual values of resistance and current may vary, depending on the manufacturer.) Each IC package contains six individual inverter circuits, all contained within a 14-pin *dual in-line package* (DIP). Since the output circuits of most 74XX logic gates are similar to either the 7404 or 7405 configuration, we will analyze these two ICs for controlling thyristor gate circuits.

The 7404 "Totem Pole" Output

As indicated earlier, a binary 1 applied to the input of an inverter produces a binary 0 at the output terminal. When the input signal is switched to a binary 0, the inverter output goes to a binary 1. The binary 1 in TTL logic is usually defined as a positive voltage level between about $+2$ and $+5$ V—the specific value depends on the source that provides the positive voltage and the loading effects of input circuits. Similarly, binary 0 is usually defined as approximately zero volts, or ground potential.

Figure 6-12A shows the action of the 7404 inverter stage. When a binary 1 is applied to the input (the emitter connection of Q1), the base-collector junction becomes forward biased. Accordingly, Q2 and Q4 are switched into saturation, and Q3 is cut off. This holds the output voltage to approximately zero. Changing the input signal to a binary 0 is equivalent to shorting the input terminal to ground. This reduces the Q1 base-collector current to zero and cuts off Q2 and Q4. Therefore, Q3 turned on, allowing the output signal to rise to a positive voltage level, or binary 1. The switching action for either transition is accomplished in less than 20 ns.

Note that the positive output voltage level produced by a binary 0 input is adequate to turn on low-power thyristors such as SCRs or triacs. The maximum output current for a shorted output terminal is approximately 30 mA. However, this approach is seldom used due to the limited power-dissipation rating within the 7404 and the electrical isolation required between the 7404 and the thyristor circuit.

The 7405 Open-Collector Output

The 7405 inverter circuit (Fig. 6-12B) is similar to that of the 7404 except that Q3 is deleted. When a binary 1 is applied to the input terminal, Q4 is saturated. Changing the input to a binary 0 cuts off Q4.

The open-collector configuration permits independent use of the output transistor in external control circuits. Theoretically, we could use this transistor to control the firing of low-power thyristors directly. However, such circuits are not practical due to the need for electrical isolation between the 7405 and the thyristor control circuits.

Interface Techniques

Fast-switching optoisolators, or optically coupled drivers, are employed in most microprocessor or digital control systems as an interface with thyristor gate-trigger circuits. Figure 6-13 shows a basic circuit employing an optoisolator between a 7405 open-collector inverter and a triac. The Motorola MOC3010 optically isolated triac driver is typical of the optoisolators available for this application. A binary 0 applied to the input circuit of the 7405 prevents the optoisolator from conducting. This condition maintains the triac in the off, or nonconducting, state. When the input to the 7405 is changed to a binary 1, the optoisolator permits a gate trigger signal to be applied to the triac. Thus, the triac is turned on whenever a binary 1 is applied to the input of the 7405 inverter stage. The snubber network, consisting of R4 and C1,

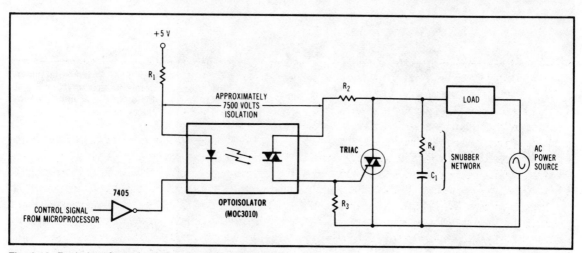

Fig. 6-13. Basic interface circuit for triac switching system.

is to prevent premature firing of the thyristor. Resistors R1 and R2 are used for current limiting. A resistive termination for the gate circuit of the triac is provided by R3—this helps to prevent false triggering action from transient signals.

No attempt was made to assign specific values to the components shown in Fig. 6-13. The selection of the triac must be based on the load current and voltage requirements. This, in turn, dictates the gate trigger signal characteristics. Although not shown in this circuit, an rfi filter may be required to prevent radiation of interfering signals into the microprocessor control system.

Additional thyristor switching characteristics may be required for specific installations. For example, zero-voltage-crossing switching for SCRs or triacs may be needed to eliminate rfi problems or to ensure precise control of power applied to the load. In this case, you may elect to use a zero-voltage-crossing detector in the gate-trigger circuit. Some commercial high-power SCR controllers are available with an optional zero-voltage-crossing detector circuit.

Commercial Microcomputer Control Systems

Many major electronic manufacturers produce and market microprocessor-based control systems of varying capabilities. Most of these control systems are available with input and output control modules that can be adapted to almost any industrial application.

Figure 6-14 shows the Motorola M68MMFLC 1/2, a commercial microcomputer control system designed for installation in industrial environments. This industrial controller, using either an 8-bit or 16-bit microprocessor, provides for real-time control of industrial processes. It is designed for front loading and has a 14-slot *motherboard* with card cage.

The microcomputer configuration is determined by the user's application. RAM requirements may range from 4K (4096) 8-bit words to 64K words or more. Special ROM may be required for some standardized applications. Isolated input and output interface modules permit connection to ac- or dc- operated sensors or loads, respectively. These modules incorporate optocouplers which provide an isolation of 3750 V_{RMS} or even higher between the microcomputer control system and the outside world. Thus, the output modules can be connected directly to the gate-trigger circuits of high-power thyristors. Similarly, input modules can

Fig. 6-14. An industrial microcomputer system (courtesy of Motorola Semiconductor Products Inc.).

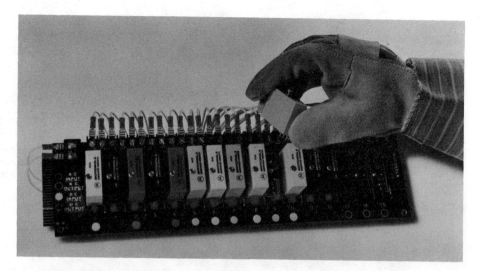

Fig. 6-15. Interface modules (courtesy of Motorola Semiconductor Products Inc.).

be used to detect zero-voltage-crossing conditions. With this capability, the microprocessor can be programmed to turn on high-power thyristors at required time intervals. Figure 6-15 shows typical input and output modules developed for use with microcomputer control systems.

Figure 6-16 is a block diagram of an industrial control system employing a zero-voltage-crossing detector. A zero-voltage-crossing detector connected to the load power source causes a signal to be applied to the input module for every zero voltage crossing. The input module couples this signal

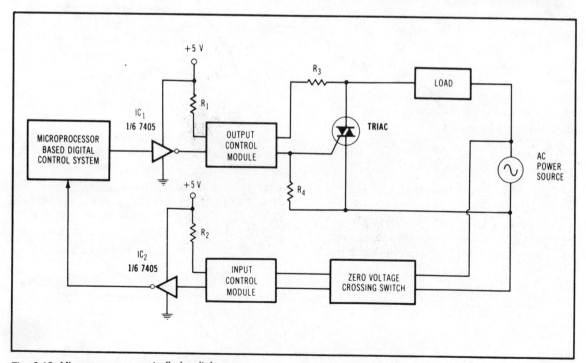

Fig. 6-16. Microprocessor-controlled switch.

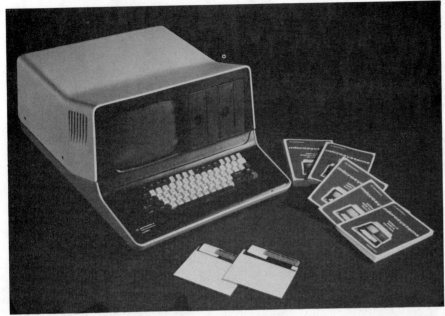

Fig. 6-17. A microcomputer software development sys- (courtesy of Motorola Semiconductor Products Inc.).

to the microprocessor-based control system via IC2. When the internal stored program within the microcomputer requires that the triac load is to be switched on, the microprocessor waits for a zero-voltage-crossing condition. As soon as this condition is sensed, the microprocessor transmits a binary 1 signal to IC1. This action causes the output module to turn on the triac. When the program controlling the process determines that the triac should be turned off, the microprocessor causes a binary 0 to be applied to the input of IC1.

Software Development

The initial development, testing, and subsequent maintenance of microprocessor programs for complex industrial control systems require the use of a *software development system*. Most manufacturers of microprocessor-based control systems will provide the user with the necessary hardware as well as programming aids and general utility programs. Figure 6-17 shows a microcomputer-development system designed for software, or *program*, support. Called the EXORset 30, this system was designed to support the Motorola 6809 and related microprocessor-based control systems. The *minifloppy* disc storage units permit permanent storage of programs developed by the user as well as programming aids provided by the microprocessor manufacturer.

Some users may elect to seek outside assistance for the initial software development phase. This permits the user to concentrate on training of programmers and maintenance personnel for the operational phase of the project. Either the microprocessor manufacturer or an independent software development company can be contracted to perform this task.

References

Cannon, Don L., and Luecke, Gerald. *Understanding Microprocessors*. Dallas, Texas: Texas Instruments Inc., 1978 (Developed and published for Radio Shack by Texas Instruments Learning Center).

Floyd, Thomas L. *Digital Logic Fundamentals*. Columbus, Ohio: Charles E. Merrill Publishing Co., 1977.

Gibson, John H. "A Computer-Controlled Light Dimmer," *Byte, the Small Systems Journal*, January, 1980, pp. 56-72, February 1980, pp. 72-80.

McWhorter, Gene. *Understanding Digital Electronics*. Dallas, Texas: Texas Instruments Inc., 1978 (Developed for Radio Shack by Texas Instruments Learning Center).

Patrick, Dale R., and Fardo, Stephen W. *Industrial Electrical Systems*. Indianapolis, Indiana: Howard W. Sams & Co., Inc., 1977.

"The 8086 Family: Helping To Solve The Software Crisis Of the 80's," *Intel Innovator*, Spring 1980. Santa Clara, California: Intel Corp.

Index

A
ac control circuits, 25
ac phase-delay switching circuits, 40
alloy, 16
atomic structure, 2
automation, 159
avalanche current, 27

B
base, 15
bidirectional triode thyristor, 71
binary numbers, 165
bipolar transistors, 12, 13
breadboards
   typical, 64
British Thermal Units, 143
BTU, 143
bytes, 166

C
calories, 143
collector, 15
commercial heat sinks, 149
conducted rfi, 152
conductivity, 3
conductors, 3

D
DeForest triode tube, 1
depletion zone, 9

diac theory of operation, 97
diffusion, 16
digital computer, 160
   first-generation, 160
   second-generation, 160
digital logic families, 171
digital logic gates, 170
diodes
   signal, 12
   tunnel, 12
   varactor, 12
   zener, 12
diodes
   rectifier, 12
discharging time constant, 113
doped material, 5
drain, 15

E
electrical current, 3
electrical isolation, 54, 145
electron, 2, 3
electron current, 3
electrons, 7
emitter, 15
epitaxial, 16

F
field-effect transistors, 15
Fleming valve, 1
forward biasing, 9

G
gate, 15
gate pulse triggering, 29
gate turn-on circuits, 38
gating characteristics, 28
grown junction, 16

H
heat sinks, 140
   commercial, 149
heat-sink analysis, 146
hexadecimal numbers, 166
holding current, 21, 24
hole current, 3

I
insulators, 3
interface techniques, 173

J
JFET, 15
junction barrier, 9

L
LASCR characteristics, 51
leakage current, 10
light-activated SCR, 51
logic gates, 170
LSI, 161

M

magnetic amplifiers, 42
majority carriers, 6
microcomputer control systems
   commercial, 174
microcomputers, 164
microprocessor instructions
   basic categories of, 169
microprocessor organization, 167
microprocessor system
   typical 8-bit, 168
microprocessors, 164
minority carriers, 6
monocrystalline, 16
MOSFET, 15

N

n-channel, 15
negative charge, 2
negative pole, 5
neutron, 3
nibbles, 166
nonrepetitive-pulse, 146
npn transistor, 13

O

ohm, 3
Ohm's law, 4
optoisolator SCR circuits, 54
optoisolator specifications, 56

P

p-channel, 15
peak inverse voltage, 12
pentavalent, 5
phase-control switching, 85
photocoupler, 54
PIV, 12
pn junctions, 7
   biased, 9
   unbiased, 8
pn-junction diode operating
   characteristics, 11
pnp transistor, 13
polycrystalline, 16
positive charge, 2
positive pole, 5
power supply
   typical, 63
processing
   automated, 159
production
   automated, 159
programmable unijunction transistor, 118, 122
proton, 3
pulsed gate triggering, 29
PUT, 118, 122
PUT applications, 121
PUT specifications, 121
PUT theory of operation, 120, 121

R

radiated rfi, 152
radio frequency interference, 150
rectifier diodes, 12
relaxation oscillator, 113
repetitive-pulse, 146
resistance, 3
reverse biasing, 10
rfi suppression, 152
robots, 159

S

saturable reactors, 42
SBS, 125, 126
SCR, 15, 21, 22, 23
   fast switching, 33
SCR ac operation, 68
SCR bridge circuit, 60
SCR circuits
   phase-control, 139
SCR controller configurations, 48
SCR devices
   specialized, 55
SCR gate characteristic curves, 37
SCR operating characteristics, 26, 29
SCR phase-control test circuit, 68
SCR power controllers
   commercial, 49
SCR specifications, 28
SCR test circuit, 66
SCR turn-on methods, 45
SCS neon-tube driver circuit, 107
semiconductor materials, 3
semiconductor theory, 1
semiconductors, 3
   n-type, 5
   p-type, 6
Shockley diode, 95
Shockley diode basic construction, 96
Shockley diode characteristic curve, 96
Shockley diode common symbols, 96
signal diodes, 12
silicon bilateral switch, 125, 131
silicon diodes, 11
silicon unilateral switch, 124, 127
silicon wafers
   manufacture of, 17
silicon-controlled rectifier, 1, 15, 21
silicon-controlled rectifiers, 30
silicon-controlled switch, 100
silicon-controlled switch
   characteristics, 103
small-scale integration, 160
software development system, 176
source, 15
SSI, 160
static switching circuits, 38

steady-state, 146
SUS, 124, 126
switching
   zero-voltage, 45
switching characteristics, 28
switching circuits
   ac phase-delay, 40
   static, 38
switching time, 21

T

thermal compounds, 145
thermal conductivity, 143
thermal resistance, 144
thermal spectrum, 141
thyristor, 15
thyristor circuits
   rfi interaction with, 154
thyristor control system, 137
thyristor devices
   testing, 158
thyristor manufacture, 16
thyristor symbols, 19, 20
thyristor testers, 158
thyristors, 159
   heat flow in, 141
   proper, 138
   typical, 18
transistor
   npn, 13
   pnp, 13
   programmable unijunction, 118, 122
   unijunction, 114
transistors, 12
   bipolar, 12, 13
   field-effect, 15
   unijunction, 108
triac, 15, 71
   basic construction, 72
   low-power, 80
   medium-power, 83
   theory of operation, 71
   typical, 72
triac ac operation, 90
triac characteristics, 87
triac commutation, 77
triac gate characteristics, 75
triac specifications, 79
triac static switching circuits, 79
triac switching time, 77
triac symbol, 72
triac turn-on methods, 79
triode ac semiconductor, 71
triode tube
   DeForest, 1
trivalent, 6
tunnel diodes, 12

U

UJT typical specifications, 112
unijunction transistor, 114
unijunction transistors, 108

V

varactor diodes, 12
VLSI, 161

voltage breakover turn-on, 73

W

words, 166

Z

zener diodes, 12
zener voltage, 12
zero-voltage switching, 45, 85

Other Bestsellers From TAB

☐ **THE MASTER SEMICONDUCTOR REPLACEMENT HANDBOOK—Listed by Industry Standard Number & THE MASTER SEMICONDUCTOR REPLACEMENT HANDBOOK—Listed by Manufacturer's Number**

This GIANT, two-volume information source gives you listing of over 160,000 semiconductor devices—transistors, diodes, and integrated circuits—and more than 600,000 manufacturer numbered parts produced by all the major manufacturers—General Electric, Radio Shack, Motorola (commercial and HEP), Philips (formerly Sylvania), and RCA. With the wealth of data supplied by these two volumes, it's a snap to locate a replacement for almost any semiconductor ever made. Whether you're a hobbyist or experimenter, professional engineer or technician, these handbooks are invaluable workbench companions . . . "tools" you'll refer to time and time again! Two Volumes, 1,356 total pp., 6" × 9" Format, Vinyl.
Price (2 Vols.) $35.00   **Book No. 5000**

☐ **55 EASY-TO-BUILD ELECTRONIC PROJECTS**

If you're looking for a gold mine of exciting, fun-to-build, *and useful* electronic projects that are both easy and inexpensive to put together . . . here's where you'll hit pay dirt!! Here are more than 50 unique electronic projects selected from the best that have been published in *Elementary Electronics Magazine* . . . projects that have been tested and proven so you know that they really work!
Paper $14.95   **Hard $21.95**
Book No. 1999

☐ **THE PERSONAL ROBOT BOOK**

This state-of-the-art "buyer's guide" fills you in on all the details for buying or building your own *and even how to interface a robot with your personal computer!* Illustrated with dozens of actual photographs, it features details on all the newest models now available on the market. Ideal for the hobbyist who wants to get more involved in robotics without getting in over his head. 192 pp., 105 illus. 7" × 10".
Paper $12.95   **Hard $21.95**
Book No. 1896

☐ **HANDBOOK OF REMOTE CONTROL & AUTOMATION TECHNIQUES—2nd Edition**

From how-to's for analyzing your control needs to coming up with the electronic and mechanical systems to do the job, the authors provide a wealth of information on just about every kind of remote control system imaginable: temperature, light, and tone sensitive devices . . . pressure or gas sensors . . . radio controlled units . . . time controlled systems . . . and microcomputer interface systems . . . even robots. Shows how to analyze your test control needs and find exactly the electronic or mechanical system to do your job. 350 pp., 306 illus.
Paper $13.95   **Hard $21.95**
Book No. 1777

☐ **ENCYCLOPEDIA OF ELECTRONICS**

Here are more than 3,000 complete articles covering many more thousands of electronics terms and applications. A must-have resource for anyone involved in any area of electronics or communications practice. From basic electronics or communications practice. From basic electronics terms to state-of-the-art digital electronics theory and applications . . . from microcomputers and laser technology to amateur radio and satellite TV you can count on finding the information you need! 1,024 pp., 1,300 illus. 8 1/2" × 11".
Hard $60.00   **Book No. 2000**

☐ **ELECTRONIC AND MICROPROCESSOR-CONTROLLED SECURITY PROJECTS**

Have the latest security systems protecting your family, home and property for a fraction of the price of installing a commercially-available system. With only a basic understanding of electronic principles, you can design, build, and install devices ranging from an electronic light controller to a complete 6502-microcomputer based security system. 240 pp., 119 illus. 7" × 10".
Paper $14.95   **Hard $21.95**
Book No. 1957

☐ **THE CET STUDY GUIDE—Wilson**

Now, for the first time, there's a CET study guide to help you prepare for the ISCET Journeyman Consumer CET examination . . . to help you review exactly what you'll need to know in terms of theory and practical applications in your specialty—consumer electronics including TV, radio, VCR, stereo, and other home electronics. Includes a 75-question practice test with answers plus indispensable hints on taking the CET exam. 294 pp., 67 illus.
Paper $11.95   **Hard $16.95**
Book No. 1791

☐ **FUNDAMENTALS OF TRANSDUCERS**

A fascinating, hands-on introduction to the world of electronic—transducers and sensor! Far more complete than any other experimenter's guide to transducers, this sourcebook will not only bring you up to speed on today's transducer and sensors technology, it will become an indispensable reference that you'll find yourself turning to time and time again for quick, to-the-point information on all types of monitoring and measuring applications. 364 pp., 229 illus.
Paper $14.95   **Hard $21.95**
Book No. 1693

Other Bestsellers From TAB

☐ **DESIGNING AND BUILDING ELECTRONIC GADGETS, WITH PROJECTS—Carr**

Chock-full of electronics gadgets and devices that really work, this is a different kind of project-building guide . . . not one that requires meticulous duplication . . . but one that offers a practical look at the "hows" and "whys" of electronic design, using a wealth of fascinating and workable applications. You'll find the expert guidance you need to make useful electronics devices and the inspiration you need to devise your specific needs and ideas! Hobbyists of all experience levels will find this guide unusually valuable! 406 pp., 214 illus.

Paper $12.95　　　　　　　　　　　　Hard $19.95
Book No. 1690

☐ **UNDERSTANDING ELECTRONICS—2nd Edition**

Even more complete than the original edition that has been a classic handbook for hobbyists in all experience levels, this new expanded version covers the most recent developments and technologies in the field as well as basic fundamentals. Even the not-so-expert hobbyist will be interested in new chapters dealing with transistor characteristics and the basic guide to selecting transistors for specific uses . . . new material on amplifiers, oscillators, power supplies, and more. It's your key to putting together circuits that really work, right from the start! 208 pp., 191 illus.

Paper $9.95　　　　　　　　　　　　Hard $15.95
Book No. 1553

☐ **HOW TO DESIGN/BUILD REMOTE CONTROL DEVICES—Stearne**

A computer can control and operate virtually anything from room environment to complex machines, if you give it the capabilities. And here is all the information you need to apply remote control to almost any situation imaginable, in a book that makes remote control as simple as it really is. Includes ready-to-build projects: a garage door opener, alarm systems, and more! 384 pp., 288 illus.

Paper $15.95　　　　　　　　　　　　Book No. 1277

☐ **THE CET EXAM BOOK—Glass and Crow**

This is a sourcebook that can help you prepare effectively for your beginning level CET exam (the Associate Level Exam). Written by two experts in the field, this is the only up-to-date handbook designed specifically to help you prepare for this professional certification test. Included are samples of all the questions you'll encounter on the exam—plus answers to these questions and explanations of the principles involved. 224 pp., 189 illus.

Paper $9.95　　　　　　　　　　　　Hard $14.95
Book No. 1670

☐ **HANDBOOK OF ADVANCED ROBOTICS—Safford**

Here's your key to learning how today's sophisticated robot machines operate, how they are controlled, what they can do, and how you can put this modern technology to work in a variety of home, hobby, and commercial applications. Plus, you'll find complete instructions for building your own remote-controlled hobby robot; get a detailed look at currently available commercial robots and androids; and gain an understanding of the many different types of machines that are classified as robots—including electrically, hydraulically, and pneumatically operated units. 480 p., 242 illus.

Paper $16.50　　　　　　　　　　　　Book No. 1421

☐ **THE COMPLETE HANDBOOK OF ROBOTICS**

How to design and build ANY kind of robot . . . including ones with microprocessor "brains"—PLUS how to interface robots with a computer! It's a single sourcebook that contains all the techniques you'll need for creating, designing, building, and operating your own robot . . . with enough options to create a whole family of robotic wonders! Covers robot circuits, controls, and sensors. 364 pp., 137 illus.

Paper $14.95　　　　　　　　　　　　Book No. 1071

*Prices subject to change without notice.

Look for these and other TAB books at your local bookstore.

TAB BOOKS Inc.
P.O. Box 40
Blue Ridge Summit, PA 17214

Send for FREE TAB catalog describing over 900 current titles in print.